Teaching Evolution in a Creation Nation

The History and Philosophy of Education Series

Edited by Randall Curren and Jonathan Zimmerman

Teaching Evolution in a Creation Nation

Adam Laats and Harvey Siegel

The University of Chicago Press CHICAGO AND LONDON

The History and Philosophy of Education Series is published in cooperation with the Association for Philosophy of Education and the History of Education Society.

ADAM LAATS is associate professor of education and history at Binghamton University, State University of New York. He is the author of *The Other School Reformers* and *Fundamentalism and Education in the Scopes Era*. HARVEY SIEGEL is professor of philosophy at the University of Miami. He is the author of several books, including *Relativism Refuted, Educating Reason,* and *Rationality Redeemed?,* as well as the editor of *Reason and Education* and *The Oxford Handbook of Philosophy of Education.*

The University of Chicago Press, Chicago 60637
The University of Chicago Press, Ltd., London

Printed in the United States of America

25 24 23 22 21 20 19 18 17 16   1 2 3 4 5

ISBN-13: 978-0-226-33127-0 (cloth)
ISBN-13: 978-0-226-33130-0 (paper)
ISBN-13: 978-0-226-33144-7 (e-book)
DOI: 10.7208/chicago/9780226331447.001.0001

Library of Congress Cataloging-in-Publication Data
Laats, Adam, author.
Teaching evolution in a creation nation / Adam Laats and Harvey Siegel.
pages cm — (History and philosophy of education ; 1)
ISBN 978-0-226-33127-0 (cloth : alk. paper) — ISBN 978-0-226-33130-0 (pbk. : alk. paper) — ISBN 978-0-226-33144-7 (e-book) 1. Evolution (Biology)—Study and teaching. 2. Creationism—Study and teaching. 3. Education—United States—Religious aspects. I. Siegel, Harvey, 1952– author. II. Title. III. Series: History and philosophy of education ; 1.
QH362.L33 2015
576.8076—dc23
2015015469

♾ This paper meets the requirements of ANSI/NISO Z39.48-1992 (Permanence of Paper).

CONTENTS

ACKNOWLEDGMENTS

We are grateful to many. We thank Randall Curren and Jon Zimmerman, co-editors of this book series, for the invitation to write this book and for their incisive editorial advice, and the other series authors for their helpful feedback on work in progress. We thank Harry Brighouse, Otávio Bueno, Peter Luyxk, Denis Phillips, Michael Veber, and an anonymous referee for helpful criticisms and suggestions. We are particularly grateful to Glenn Branch and Mike U. Smith for detailed comments on the entire manuscript. One of us (HS) gratefully acknowledges the support of the Spencer Foundation, which facilitated the writing of chapter 8.

INTRODUCTION

The Evolution of an Educational Controversy

What do you *know* about evolution? What do you *believe* about it? If you are like many Americans, these questions are not as simple as they might seem. Indeed, in a controversial move, the National Science Board (NSB) eliminated data about evolution from its 2010 report about Americans' scientific literacy.[1] Why? Because in the case of evolution, knowledge and belief have become inextricably intertwined. What Americans say we "know" has become hopelessly muddled with what we "believe." And nowhere is this more muddled than in the public schools.

To pick apart this tangled knot, this book examines both the historical and the philosophical issues at the heart of these controversies in the United States.[2] From a historical perspective, controversy concerning evolution must be understood as part of a long tradition in American education. Those who choose not to believe or accept[3] evolutionary theory; those who are unable or unwilling to accept it; those who fight to keep their children insulated from such beliefs: all these evolution opponents must be seen as part of the tradition of religious dissent in American education.[4]

Evolution education needs a clarification of terms. Just as the NSB has done, so thoughtful observers of these controversies must notice the need to separate questions of knowledge from those of belief. It is not the purpose of public schools to impose religious beliefs on students, but it is precisely the purpose of such schools to expose students to the best knowledge. Evolutionary theory involves both. Students must understand evolutionary theory and know a lot about its main features and the evidence that supports it. Whatever religiously motivated dissenters may say, evolutionary theory is a fundamental building block of knowledge in the modern world. But this fact does not imply that dissenting students should be expected to believe that evolutionary theory is true, given their religious beliefs. For in spite of what science-motivated pundits may say, evolutionary theory has religious implications for many students. While we might hope that all science students come to believe it, it is not the business of public schools to enforce or require belief. Many students will, upon achieving an understanding of evolutionary theory and the evidence that supports it, come to believe it. When students do not believe it, whether for religious or other reasons, public schools should not insist on belief. Rather, student knowledge and understanding of evolution are the marks of successful evolution education.

In short, we need to approach the questions raised by evolution education as we would any other questions about education. The relevant questions are not merely those raised by the nature of science,[5] or the politics of schooling,[6] or the nature of social movements.[7] The questions centrally involve issues of teaching and learning. As with any topic, teachers of evolutionary theory must vigorously cultivate respect for all students. If students come from backgrounds that dissent on religious or cultural grounds, that dissent must be acknowledged, but even religious dissenters and students from cultural minority groups must not be excused from learning about the ideas and issues fundamental to modern life.

Before we wade into the history of these conflicts and the philosophical issues that arise, we need to clarify our terms. When we speak about evolution, religion, science, and education, we risk getting lost in the weeds of competing terms and definitions. It might be tempting to call one side science and the other religion, but that only confuses the issue, since some people on both sides claim to be the defenders of both true science and true religion.[8]

A more practical name for the first of these competitors might be "evolution supporters." Over the course of the twentieth century, this group solidified its control of the institutions of mainstream science: research universities, professional organizations, and peer-reviewed science publications. Evolution supporters have worked to increase the amount and quality of what they consider legitimate evolution education in America's schools. Too often, in their view, science education has been unduly influenced by erroneous popular beliefs. They have warned that such acquiescence to popular but scientifically suspect beliefs will put the United States in a woefully noncompetitive position.

The supporters of evolution education have usually based their authority on the findings of modern science. They find encouragement in the fact that nearly every mainstream scientist agrees with their fundamental claims concerning evolution. Like Galileo, they believe that whatever wrongheaded unscientific critics might say, "it still moves." Evolution is as much a fact of nature as gravity. There may be vigorous disagreement about the patterns and processes of evolution, they believe, but nearly all scientists will agree with the notion that life on earth has evolved in a series of modifications from earlier forms. These evolution supporters often chafe at the popularity of what they consider "nonscience," "dead science,"[9] or, as one astute observer has called it, "zombie science."[10] They call their opponents ill-informed and poorly educated. In more heated moments, they may call them ignorant witch hunters or medieval flat-earthers.

The other side in this ongoing cultural conflict could be called the "evolution opponents." Over the course of the twentieth century, they have watched in dismay as their mainstream scientific credibility has been jerked roughly away, but that has not discouraged their claims to the mantle of true science. It has not slowed down their campaigns to ensure that the theory of evolution is defanged in schools. To that end, evolution opponents have built an array of their own institutions to support and promulgate their vision of the origins of Earth and humanity. Most often, in the United States, that vision is based on religious beliefs, usually—but certainly not always—conservative Protestantism. A vibrant network of anti-evolution colleges, K–12 schools, and research organizations has been bolstered by America's traditionally robust network of churches and religious denominations.

Over the course of the twentieth century, to the never-ending surprise of many evolution supporters, this fractious collection of evolution opponents has become more, not less vigorous in its opposition to mainstream science. Much of the opposition has moved to a more radical position on the origins of the planet and its life-forms. It has become more consistently dedicated to the idea of a young earth and a worldwide flood, which it takes to be the proper interpretation of scripture, especially the Bible's book of Genesis.[11] And the anti-evolution coalition has retained its influence on American popular beliefs about life's origins. Gallup polls between the 1980s and the present consistently have found that almost 50 percent of American adults agreed with the notion that "God created human beings pretty much in their present form at one time within the last 10,000 years or so." Contrary to the expectations of some evolution supporters, these evolution skeptics could not simply be dismissed as uneducated and unaware of modern science. As political scientists Michael Berkman and Eric Plutzer concluded recently, "Opposition to evolution is not simply a specific instance of scientific illiteracy. Rather, it appears that anti-evolutionists *choose* not to accept evolution."[12]

For whatever reason evolution opponents dismiss evolutionary theory, they have often been just as contentious and just as politically active as have evolution supporters. Just as evolution supporters have accused their opponents of small-minded obscurantism, so evolution opponents have at times insisted that evolution supporters are at best fooled by the glamour of false science and are at worst bitter atheists, hating all traditional values including those of family and country.[13]

With such accusations as part of the script of public disputes over the teaching of evolution, it is not surprising that these controversies have been fractious and angry. As the historical chapters in this volume will describe, public controversies since the early decades of the twentieth century have often degenerated into mere name-calling. Each side has accused the other of believing in an insupportable absurdity. For example, after state lawmakers in Kentucky almost passed the nation's first anti-evolution law in 1922, one Kentucky evolution supporter called their deliberations "unwise, absurd, ridiculous."[14] Another evolution supporter in Kentucky's legislature lamented that if they had banned evolution, there "would be little left for [schools] to teach."[15]

Their opponents did not agree with them on much, but they agreed that their opposition had lost touch with reality. One 1920s anti-evolutionist insisted that only "lunatics" could support the teaching of evolution to young minds.[16] One of the most prominent evolution opponents of that decade, William Jennings Bryan, denounced the scientific pretensions of evolution supporters as merely "laughable." He insisted that he did not have any complaint against ignorance and absurdity, except when those things tried to pass themselves off as legitimate science.[17]

In part, these accusations reflected each side's sincere inability to understand the other. The proofs for creation by divine plan or for evolution from earlier forms both seemed so profoundly convincing to their supporters that any disagreement seemed irrational, if not criminal. But these arguments about the absurdity of the opposition have been more than logical arguments about science; they have included more sweeping cultural considerations. The story of the struggle over evolution education has been a record of such wide-ranging claims, an irreducible stew of scientific, legal, political, regional, religious, demographic, historical, racial, and pedagogical ideas. Attempts to understand it as purely a dispute between science and religion or between educated and uneducated factions all founder in the face of the historical record.

The history of conflict over evolution education has shown this to be a durable tension in American society. It must be understood as such, rather than as a series of isolated courtroom showdowns. Yet largely because of the segregation of the two cultural visions at stake, each legal battle since the Scopes trial of 1925 has been anticipated by both sides as the final word, the decider of the issue once and for all. Every time, the losing side pooh-poohs the decision, while the winner trumpets it. Every time, activists express their shock and surprise when the issue of evolution education again becomes a prominent public controversy. Each new showdown is predictably labeled "Scopes II." Yet although the overarching issue at the center of these dramatic conflicts has remained evolution education, the nature of the debate has changed dramatically over the course of the twentieth century. The historical chapters in this volume will attempt to describe those changes. They will also move beyond and behind these singular episodes to expose the wider cultural conflicts behind them.

From the first publication of Charles Darwin's *On the Origin of Species by Means of Natural Selection* in 1859, scientists and the wider public debated its premises and conclusions. As chapter 1 argues, these early discussions are best understood by looking at broader cultural changes in the nature of learning and higher education. In chapter 2, we look at the outbreak of controversy over the issue of teaching evolution in America's public schools in the 1920s. Too often, histories of evolution and creationism focus only on such controversies. In order to make sense of the long conflict, however, as we see in chapter 3, we need to spend at least as much effort understanding important cultural changes that took place away from the headlines. Only then can we understand the new round of evolution battles that erupted in the 1960s. As chapter 4 argues, evolution opponents in the later half of the twentieth century often based their case on their rights as persecuted minorities.

Throughout this history, though positions and personalities have changed dramatically, opposition to the teaching of evolution that springs not from eccentric or idiosyncratic individuals but from a sustained protest by a large cultural minority group has remained constant, in keeping with a long tradition of popular religious dissent. The status of evolution opposition as a minority position advocated by a dissenting group raises key philosophical issues concerning the nature of science and the purposes of education. Chapter 5 treats the most basic issues: What is a scientific theory? Can science be cleanly distinguished from nonscience? What is the difference between science and religion? Are either evolutionary theory or "creation science" genuine science? Is either (or both) rather merely thinly veiled religion? What is the state's obligation with respect to biology education? Must it be "neutral"? If so, what does such neutrality require? Chapter 6 addresses the specific issues raised by intelligent design, the successor theory to creation science. Chapter 7 offers an account of the aims of science education in general and evolution education in particular. We defend the view that biology teachers need not aim to foster student *belief* in evolution, but rather should cultivate student *knowledge* and *understanding* of the theory and its supporting evidence. In particular, we argue that biology students should be helped to understand that evolutionary theory provides the best extant *scientific* account of its domain. Whether or not they actually believe it is a function of many factors—religious, cultural, and psychological—and the science teacher will have done her job well if such student understanding is achieved. In chapter 8, we consider the issues raised by thinking of the conflict in cultural terms.

While we think that it is indeed sensible and important to understand the conflict in these terms, doing so does not materially affect curricular matters. For instance, we are not recommending any specific reform to biology curricula in public schools. Rather, by understanding the controversy as an episode in the long history of religious and cultural dissent in US public education, we hope to suggest new ways of overcoming it. Such ways respect both evolutionary theory and religious or cultural objections to it. By doing so, we show the path forward to a time in which all parties to the dispute enjoy the legitimacy to which they are entitled.

CHAPTER 1

Higher Education and a New Culture of Science

Edward Birge was tired. The winter of 1921 had been a long one, even by the frigid standards of Madison, Wisconsin. After over forty years of life as a professor and administrator at the state university, he could handle the weather. However, in addition to the cold this year, in addition to his usual duties as university president, he had endured long months of relentless public attack. He was tired of it, but not nearly ready to surrender. He did not see the need.

But he did want to share his troubles. In the early spring of 1922, he wrote in commiseration to fellow academic scientist Edwin G. Conklin. Conklin, a prominent Princeton biologist, crusaded for the scientific teaching of evolutionary theory. Good luck, Birge wished him, but beware of shaking that hornet's nest. Once you have engaged the foes of evolution, Birge warned, you will certainly be endlessly assailed, personally and vindictively. He noted from experience that at the very least, Conklin would soon be deluged with "an enormous number of letters and much fool printed stuff."[1]

The two men and their campaigns in favor of the teaching of evolutionary theory represented the new mainstream of academic life at leading institutions of higher education. Both were scientists, and both expected universities to welcome research and teaching that did not truckle to religious or cultural orthodoxies. Such attitudes represented the rising trend in American science and higher education, but in the 1920s these ideas about the proper role of higher education were still novel to the majority of American citizens. A sizable segment of the population still imagined college to be a place that would instill traditional Protestant values, not challenge them. They expected science to be the handmaiden of revealed truth, rather than to engage in a self-consciously iconoclastic search for some other sort of fundamental truth. In philosophical terms, as we'll explore in chapter 8, we might describe this as a cultural divide between those university-based research scientists who had accepted the culture of Western Modern Science and significant numbers of the broader population who had not. Although both Birge and Conklin could take comfort in the fact that they had secure berths in such schools as Wisconsin and Princeton, they could not glibly assume that the folk outside their ivory towers would defer to their scientific expertise.

When Birge sat down to write to Conklin, he knew he'd find a sympathetic ear. After his long winter, he could use some sympathy. To his surprise, he had endured ferocious hostility from critics outside the circle of academics, liberal Protestants, and scientists. By the early spring of 1922, he knew that many of his detractors assumed that he fit their stereotype of the atheistic college president who uses science to turn young minds away from traditional Christianity. But that was not true of Birge at all. Though a prominent zoologist, he was also a devout Christian, a deacon in his church, and a man who had led Bible classes for twenty-five years.[2]

He hadn't looked for a fight with evolution opponents, but a fight had found him nonetheless. After a speech delivered in May of 1921 at the Madison campus by leading evolution foe William Jennings Bryan, Birge penned a long open letter denouncing what he saw as Bryan's dangerously misleading argument. Bryan had argued that the teaching of evolution at Wisconsin could have no result but atheism. He had denounced lecturers that attacked the Bible as nothing but a collection of myths. The choice, he had thundered, was between religion and morality on one hand, and evolution and sinister atheism on the other.[3]

Birge could not restrain himself. He accused Bryan himself of promoting atheism. He believed that Bryan had asserted a false division between evolution and religion. Rather, Birge argued, science and religion were both "equally divine revelations." Though he noted that some conservative theologians in the past had found science "atheistic," he disagreed. Science and the Bible, Birge insisted, "are only the same thing stated from different points of view."[4]

Birge hadn't foreseen how much of a firestorm he had stepped into. Bryan quickly gathered local support, including that of Herbert C. Noonan, a Catholic priest and president of Milwaukee's Marquette University.[5] The Wisconsin Council of Catholic Women censured Birge and pledged itself to "a vigorous campaign to eliminate false teachings at Madison."[6] Angry Wisconsinites demanded his resignation.[7]

Bryan seemed to welcome the attention. He certainly was no stranger to public controversy. His long career in the political limelight had begun with his rousing "Cross of Gold" speech of 1896. He had run unsuccessfully for president three times and resigned dramatically from his post as secretary of state over president Woodrow Wilson's aggressive response to the sinking of the *Lusitania*. He had entered the campaign against evolution with his typical energy and zeal, and the controversy against Birge allowed him the opportunity to test his arguments in public.[8]

Bryan blasted the University of Wisconsin as a typically out-of-touch university. He warned that such schools presented immediate dangers to the youth of America. "If the University of Wisconsin," he demanded,

> is to discard the Bible and substitute the guesses of scientists in its place—as it does in teaching Darwinism—and then objects to the students hearing the other side, it should . . . issue an announcement like this: Our class rooms furnish an arena in which a brutish doctrine tears to pieces the religious faith of young men and young women; parents of the children are cordially invited to witness the spectacle. Why should Christian taxpayers permit the Bible to be attacked by their hired servants where defense is not permitted?[9]

Birge lost his cool. He called Bryan "crazy."[10] Bryan called him "an autocrat."[11] Bryan even offered to pay one hundred dollars if Birge would sign a statement acknowledging his descent from an ape.[12] Birge received piles of angry mail. Letters to the editor in local papers accused him of being both weak-willed and autocratic, both unscientific and un-Christian.[13]

In the end, however, the controversy petered away, leaving Birge with stacks of hostile letters and doubtless a few headaches, but without changing his ability to govern the University of Wisconsin as he saw fit. He never had to change the university's hiring or supervision policies for professors. Although he had made an indignant public statement of his deep Protestant faith, he hadn't been forced to foster that faith as part of his presidential duties. Rather, he continued to run the university as a recognizably modern research university.

That style of higher education administration was still something of a novelty in 1922, but since the end of the Civil War it had gradually become the norm at leading colleges and universities. This slow revolution fostered a new academic climate vital to the development of controversies over the teaching of evolution. The transformed climate offered America's small but growing population of research scientists a more congenial home for research that might remove them from the orbit of popular acceptance or approval. It obviated their need to check their research findings against the tenets of accepted orthodoxies, religious or political. It created a new standard of intellectual respectability, one that measured its worth against the standards of elite international research scientists, not the traditional religious or moral beliefs of the nation.

Birge's position as a defender of academic freedom, of the university as a home for unfettered research, would have been fairly radical when he began his career at the University of Wisconsin as a lecturer in natural history in 1875. More typical at that date was the reminiscence of political scientist John W. Burgess, who remembered professors' attitudes from that same year at Amherst College. Burgess recalled that "the older minds of the faculty" still considered their job to be that of disseminating the truth. Such truth, in their opinions, had "already been found. . . . in the Bible." These older scholars, during the late decades of the nineteenth century, "regarded research as more or less heretical."[14] As Bryan argued a generation later, the prevailing vision of higher education until the late nineteenth century was that it ought to train young men, and a few women, in the precepts of life as Christian gentlemen and ladies.

As president Noah Porter of Yale University expressed it in his baccalaureate sermon of 1882, "Unbelief in some of its aspects was never more imposing to men of culture than it is at the present time. . . . Our business is to defend the truth as we believe it and to defend it because we know its worth and its power."[15] Bryan would have agreed enthusiastically. But Porter was something of a curmudgeon among college presidents at leading institutions. He knew he was fighting a rearguard action in his efforts to assert a role for the university as a defender of accepted truths and norms.

That vision of the mission of higher education was changing fast. By the turn of the twentieth century, in the words of historian of higher education Laurence Veysey, leading American universities would become "all but unrecognizable" in comparison with this older vision of proper higher education.[16] Historian Roger Geiger has identified this dramatic change as coming in fits and spurts between the years 1870 and 1920. By the end of the nineteenth century, Geiger concluded, leading research universities had committed themselves firmly to a new vision: not the training of Christian gentlemen but the deliberate pursuit of "cognitive rationality—knowing through the exercise of reason."[17] The debates over the nature of science—issues we'll discuss in more depth in chapter 5—must be seen as part of this history. Scientists didn't work in a vacuum. Rather, changing norms about the nature of science and of scientific inquiry went hand in hand with changing norms at leading research universities.

The new breed of university leader often pursued the new goal of unfettered scientific inquiry with the missionary zeal reflected in their evangelical Protestant roots. J. M. Coulter, president in 1894 of Lake Forest University in Illinois, described the "work of a university" as a great "mission field." The mission, according to Coulter, was no longer to win souls to salvation, but rather to further "the emancipation of thought . . . [previously] fettered by ignorance or superstition," a "crusade against ignorance."[18]

But by the 1920s, presidents at major research universities were driven by other factors besides their zeal to promulgate new visions of knowing and learning. They competed with one another and with leading European universities to attract and retain leading scientific researchers. Any whiff of provincialism or ideological constraint might cause a university to lose out to a more agreeable research home. For instance, the University of Michigan struggled for years to develop a world-class theoretical physics department. After only two years in his position, however, its first young faculty scientist left. He returned to Copenhagen to avoid the intellectual isolation of Ann Arbor. Physics department chairperson Harrison Randall had to consider the norms of elite international research science as he tried to build his department, not the norms of American popular beliefs about science and religion.[19]

As Veysey has demonstrated, American collegiate faculty and administration yearned at the turn of the twentieth century for "an equality with . . . Europe."[20] Researchers at leading schools had often received some or all of their graduate education at European—especially German—universities. It was only in the last years of the nineteenth century that American universities began offering a significant number of graduate degrees. Although Johns Hopkins University opened in Baltimore in 1876 with a mission to offer German-style graduate training, most other research universities, in the words of historian John Thelin, "fell drastically far short of its example until well into the twentieth century."[21]

Such experiences in internationalized graduate education led the leaders of elite universities to give more weight to the opinions and attitudes of other prominent scientists and researchers than to those of the American populace. University leaders' reliance on elite opinion exposed them to charges of unseemly attraction to foreign ideas. By the 1920s, conservative Protestant leaders had noticed the changing trends in higher education. They worried, as Bryan did, that universities had become dangerous to the faith of Protestant students. As university leaders worked to bring their research departments up to international standards, conservatives accused them of importing pernicious ideas and theories.

William Bell Riley, soon to be a leader of the fundamentalist crusade for control of several Protestant denominations, accused college leaders in 1920 of dooming their students to struggle under an "avalanche of German rationalism."[22] Anti-evolution evangelist T. T. Martin likewise located the problem with the changing nature of higher education in its pandering to German ideas. "We sent our young men to the great German universities," he argued in 1923, "and, when they came back, saturated with Evolution, we made them Presidents and head-professors of our colleges and great universities."[23] More colorfully, another fundamentalist leader warned in 1921 that higher education in once reliably religious schools had gone wrong when students "got the forty-second echo of some beer-guzzling German Professor of Rationalism."[24]

These conservative critics, though often more concerned with theology than science, had noticed an important change in the nature of higher education and the culture of science. By 1920, leading research universities had undergone a dramatic transformation. They had become more attuned to international standards of research and education than to traditional American norms of collegiate education. They had become more interested in the creation of elite research faculties than with the transmission of orthodoxy. University presidents like Edward Birge viewed their role as protecting the rights of their faculty to pursue research, rather than ensuring doctrinaire instruction.

This created a far more comfortable home for faculty to pursue research. Academic scientists moved in professional circles that allowed them to accept controversial ideas such as evolution without pausing to consider the popular reputation of those ideas. However, those scientists who chose to step outside the charmed circle of elite research university life, such as Birge and his correspondent Edwin G. Conklin, quickly discovered that evolution continued to provoke intense controversy.

Conklin's career was not too dissimilar from Birge's. Like Birge, he had been raised in a religious Protestant family. At the age of thirteen, he publicly proclaimed his faith at a Methodist revival. He impressed his family and friends with his ability to recite long passages of the Bible from memory. He even passed an examination to become a lay minister in the Methodist Church.[25]

In his career as a biologist at Princeton University, Conklin applied the evangelical zeal of his youth to promulgating his vision of scientific truth. In 1920, he articulated a vision of the meaning of science that captured the mainstream scientific understanding so coherently that it is worth quoting at some length. "The *spirit* of science," Conklin argued,

> is freedom to seek and to find truth, freedom to hold and to teach any view for which there is rational evidence, recognition that natural knowledge is incomplete and subject to revision, and that there is no legitimate compulsion in science except the compulsion of evidence.
>
> The *method* of science is to proceed from observations to tentative explanations which are then tested by further observations and experiments, thus reaching general observations or theories. Scientific theories are not mere guesses but are based upon careful, detailed observations. . . .
>
> The *aim* of real science, as well as of true religion, is to know the truth, confident that even unwelcome truth is better than cherished error, that the welfare of the human race depends upon the extension and diffusion of knowledge among men, and that truth alone can make us free.[26]

To Conklin, this vision of science meant that every traditional belief must be "critically examined." Such beliefs did not necessarily need to be jettisoned, but they must offer evidence of their continuing epistemic worthiness. Nothing less than "the future progress of mankind" depended on it.[27]

A decade earlier, he had told the American Society of Naturalists that "the fact of evolution, is by common consent, no longer a problem." Only the details of the mechanism of that evolution still needed to be worked out.[28] The scientists in his audience may not have objected. But Conklin knew that outside learned societies, there was much greater reluctance to embrace the ideas and implications of evolutionary theory. He compared such delusion to medieval belief in witchcraft, and predicted confidently that "in the clear, cold light of reason [such beliefs] fade away as all ghosts do."[29]

Conklin wrote, traveled, and lectured indefatigably to bring this vision of the true science of evolution to nonscientists. He hoped he could offer "popular" explanations of some of the complicated but convincing evidence for evolution from the field of embryology.[30] For him, this crusade needed more than just a dispassionate explanation of the facts. It needed passion, heat, even anger. When he crossed swords with evolution opponent William Jennings Bryan, Conklin wasn't content to play the role of disinterested man of science. He did not merely correct Bryan's science but lambasted Bryan as a man transformed into an ass, leading an "inquisition" based on "medieval theology."[31]

Despite his heated rhetoric, Conklin failed to win over his entire audience. Bryan's vision of science and the dangers of evolution remained popular. But although Conklin and Birge insisted that their evolutionary universe still had room for religion, Bryan refused to allow such mainstream scientists to monopolize the world of science. Though he presented himself as a political and religious leader (for example, he famously claimed that "it is more important that you trust the Rock of Ages than that you know the age of the rocks,"[32] and that "there is more science in the twenty-fourth verse of the first chapter of Genesis . . . than in all that Darwin wrote"[33]), Bryan also cultivated an image of himself as the defender of "real science."

In 1924, some time after he had joined the fray against the teaching of evolution, Bryan joined and maintained a membership in the American Association for the Advancement of Science, a decidedly pro-evolution group. But he refused to concede that the advancement of science necessarily included the acceptance of the idea of evolution.[34] He worked assiduously to promote an image of himself as a defender of science, properly understood. And, Bryan argued, those who defended and promoted the teaching of evolutionary theory were only "pseudo-scientists."[35] In a high-profile article about the growing evolution controversy commissioned by the *New York Times*, he insisted that his vision was not opposed to science. "We do not ask," Bryan declared, "for the exclusion of any scientific truth, but we do protest against an atheist teacher being allowed to blow his guesses in the face of the student."[36]

Genuine science, Bryan maintained, did not match Conklin's definition of directionless exploration in search of an unrestricted truth. Rather, Bryan relied on a different conception of science, one that had been dominant in American intellectual life until the end of the nineteenth century.[37] Science meant "classified knowledge," the organizing and sorting of verified facts according to a proven authority.[38] That authority, for Bryan, included the scriptures. Facts that conflicted with the proven authority must be viewed skeptically and ultimately rejected.

As Bryan correctly pointed out, "Tested by this definition, Darwinism is not science at all; it is guesses strung together."[39] He was right. If Darwin's theory of evolution had been tested by that definition, it would not have qualified as science. But that definition of science no longer had any adherents among mainstream research scientists. Similarly, Bryan stated that "Natural Selection is being increasingly discarded by scientists."[40] He was right again. However, his implication that mainstream scientists had discarded the theory of evolution was not true.

At the time Bryan wrote, leading American natural scientists had universally accepted the notion that life had evolved from previous forms.[41] But that didn't mean that all scientists embraced Darwin's hypothesis, or that they agreed on the precise meaning of *evolution*. As opponents of evolution education would gleefully point out for generations, many prominent scientists disparaged Darwin's proposed natural-selection mechanism. Others maintained a belief in some acts of special creation, even while accepting that most species descended with modifications from earlier species.[42]

Thus, in 1903, prominent German botanist Eberhard Dennert declared that the scientific community was "standing by the death-bed of Darwinism." Other leading scientists agreed.[43] But they only meant that the mechanism of natural selection could not possibly have accounted for the evolution of life's many forms. For mainstream scientists at the turn of the twentieth century, the idea that species had evolved wasn't in question. But it would take another thirty years for them to agree that Darwin's idea of natural selection, combined with Mendel's discovery of dominant and recessive forms (alleles) of genes, must be the main way that new species emerge.

Darwin's explosive idea operated with a longer fuse outside the natural sciences. After an immediate flurry of reviews and debate, many theologians assumed in the decade after the publication of *On the Origin of Species* that scientists would soon disprove Darwin's idea of the transmutation of species. By the 1870s, however, as the trend among scientists became clear, Protestant intellectuals bitterly disputed the implications of evolution theory, especially the idea of evolution unguided by divine will.[44] Some theologians took a strong stand against the idea of such materialistic evolution. Conservative Presbyterian theologian Charles Hodge minced no words in his 1874 analysis of the nature of evolutionary theory: "What is Darwinism? It is atheism." He noted that not all evolutionists were atheists, but argued that the idea of unguided evolution was "atheistic." It promoted the idea of a Godless cosmos.[45]

More liberal Protestant thinkers greeted the idea with greater enthusiasm. In 1885, prominent Congregationalist minister and speaker Henry Ward Beecher claimed to have "hailed the Evolutionary Philosophy with joy."[46] He argued that the inclusion of this great truth would regenerate theology and bring it "nearer to the spirit and form of Christ's own teachings."[47] Similarly, James McCosh, longtime president of Princeton University, declared in 1890 that he found "nothing atheistic" in Darwin's evolutionary theory "if properly understood."[48] He believed that Darwinism, "like every other part of God's work, would illustrate his existence and his wisdom."[49] There was nothing for Christians to fear in evolutionary thinking, McCosh insisted. "God," he wrote, "acts everywhere in nature through means." McCosh believed that evolution was God's method.[50]

In the decades between the publication of Darwin's groundbreaking book and the widespread public controversies over the teaching of evolution in the 1920s, theologians such as McCosh and Beecher contributed to the development of a broad network of influential evolution supporters. Their arguments offered an intellectually respectable path for Christians to accept evolutionary theory without abandoning their faith. By the 1920s, that network had successfully established itself as the intellectual mainstream in American life. Its leaders, like Conklin and Birge, could reassure themselves that they represented true learning, in spite of any popular opposition to their work.

That opposition also became increasingly powerful. Though there had been a transformation in elite intellectual culture between 1860 and 1920, no such revolution had changed popular ideas about the relationship between evolutionary theory and traditional faith. The notion that Darwinism promoted atheism remained influential, despite the earnest efforts of evolution supporters such as Beecher, McCosh, and Conklin.

By the 1920s, when the dominant status of evolutionary theory had led to its inclusion in science curricula for students of all ages, this notion and others contributed to a popular backlash. Pundits including William Jennings Bryan insisted that such teaching be banned from schools. A number of lawmakers bowed to the pressure and introduced appropriate legislation. When one such law passed in Tennessee, the stage was set for a conflict that would attract the world's attention to the issue of evolution education in that state's small town of Dayton.

CHAPTER 2

Evolution Education in a Jazz Age

Every history of evolution education gives pride of place to the Scopes trial. It makes sense. The media-soaked affair in the hot summer of 1925 in the small town of Dayton, Tennessee, set the stage for a century of conflict over the issue of evolution in schools. And in the decades since then, nearly every new legal case about evolution has been labeled "Scopes II."

It seems ironic, then, that the dispute in the Scopes trial was, legally speaking, not about evolution. It wasn't allowed to be. Thanks to a new body of historical scholarship about the case, most notably Edward Larson's Pulitzer Prize–winning *Summer for the Gods*, we know that the trial itself was more of a media stunt than a religious persecution or scientific showdown.[1]

Though the prosecution eventually persuaded the judge to push issues of Christianity and evolution out of the decision, observers at the time hoped for a more momentous ruling. When celebrity Bible defender William Jennings Bryan volunteered to serve for the prosecution, for instance, the trial quickly came to be seen as a showdown between evolution supporters and opponents. Bryan came to Dayton confident that his team could unmask the unscientific nature of evolutionary theory. He promised his fellow prosecutors that he could bring "many of the leading scientists" to Dayton to prove their case.[2] He assured them that his group of top scientists would prove to the world "that our side was prepared to hold its own against [the defense team's] committee of scientists."[3] Up until the trial itself, Bryan continued to believe that the Dayton trial would give him a chance to demonstrate the scientific emptiness of evolutionary theory. As he wrote privately to one of his prospective experts,

"I am expecting a tremendous reaction as a result of the information which will go out from Dayton."[4]

Bryan wasn't alone. Leading evolution opponents at the time shared his belief. Texas Baptist fundamentalist leader J. Frank Norris assured Bryan in the days leading up to the Scopes trial that he could put together a panel of "real scientists" who "could meet any hoax or fraud that might be made by the defense."[5] Like Bryan, Norris knew that his beliefs about evolution went against those of mainstream science. However, like many evolution opponents in the 1920s, he assumed that mainstream scientists had been blinded to scientific truth by their philosophical commitment to materialism, by their irrational commitment to the notion that the supernatural could have played no part in the creation of life. Fundamentalist readers of leading magazines such the *Moody Bible Institute Monthly*, from Chicago's Moody Bible Institute, or *Christian Fundamentals in School and Church*, from Minneapolis, read in the 1920s that the most "Modern Scientific Discoveries"[6] could prove that evolution would soon be discarded by all scientists, due to "recent discoveries in geology."[7] The leading scientific expert of the anti-evolution movement in the 1920s, self-educated geologist George McCready Price, assured Bryan privately that scientific supporters of evolution education were "out of date,—behind the times,—and don't know it."[8]

Bryan and other evolution opponents could not help but be taken aback when their dream team of anti-evolution expert scientists failed to materialize. They had to wince and acknowledge the bitter truth of celebrity defense attorney Clarence Darrow's taunts. When Bryan cited Price as a scientific expert in his Dayton testimony, Darrow pounced. Price, Darrow announced, was only a "mountebank and a pretender and not a geologist at all."[9] To their chagrin, evolution opponents had to acknowledge that their scientific experts had no standing in the eyes of mainstream science. That was a far cry from admitting that they weren't correct about the pernicious folly of evolution, but it called for a quick change in legal strategy.

Unbeknownst to the theologians and mainstream scientists who showed up in Dayton ready to testify for the defense that the teaching of evolution was both scientifically necessary and theologically safe, the prosecution had agreed to revise its approach. Instead of fighting toe to toe on the issue of evolution, it decided to focus on a much narrower strategy of simply arguing that John Scopes had broken a law of Tennessee. It agreed to argue that evolution wasn't on trial. Only Scopes was.[10]

The defense must have been disappointed. The more aggressive members of its team had also hoped to use the trial, but with a different purpose. Some of Scopes's advocates saw the event as a chance to expose the intellectual inadequacies of the anti-evolution movement. The trial had been originally sponsored by the American Civil Liberties Union (ACLU), which had offered to pay the court costs of anyone willing to serve as a test case. After Tennessee passed an act prohibiting the teaching of evolution in 1925, the ACLU hoped a trial could discourage other states from passing similar restrictions on the teaching of evolution.[11] Nevertheless, when Darrow examined Bryan on the witness stand, he, too, avoided the subject of evolution.

In many ways, Darrow was the perfect foil for Bryan. Both men had had long and brilliant public careers. Yet, unlike Bryan's reputation as a moralist and Protestant traditionalist, Darrow had long portrayed himself as a sort of village atheist on the national stage. He delighted in exposing what he saw as the cruel shortcomings of traditional faith and thinking.[12] When Bryan volunteered to serve on the prosecution team, Darrow jumped in to aid the defense. He wanted to promote evolution education as ardently as Bryan wanted to prevent it. But when he got a chance to confront Bryan, Darrow never asked him directly about evolution. To do so, the canny Darrow knew, would have invited one of Bryan's polished and convincing stump speeches on the subject. Instead, he limited his questions to traditional biblical stumpers such as the genealogy of Cain's wife. Darrow hoped to maneuver Bryan into an admission that Bryan, too, used some interpretation when he read his Bible. If even Bryan did so, fellow defense attorney Arthur Garfield Hays attested, "he must have agreed that others have the same right."[13]

These strategic decisions by both sides at the Scopes trial—and by trial judge John Raulston—ensured that the actual issue of evolution education never became a legal part of the proceedings. In spite of the media frenzy—most of which assumed that the trial would decide the propriety of evolution education once and for all—the issue itself was excluded. Yet the strategies used by the prosecution and the defense demonstrated the ultimate import of debates over evolution education in the 1920s. Like Bryan, thoughtful opponents of evolution education realized that their side could no longer summon a body of scientific experts that could claim mainstream credentials. As a result, they shifted to a strategy that relied on their remaining base of support: popular opinion. Also like Bryan, many evolution opponents quickly realized that they couldn't convince mainstream scientists about the dangers of evolution, but they could win an argument that scientists must not be allowed to cram dangerous and false ideas down people's throats. Throughout most of these debates in the 1920s, Bryan and his allies assumed that teaching evolution meant teaching atheism. As we'll argue in chapter 7, it would change the entire debate—in the 1920s and today—if both sides could separate the educational importance of understanding evolution from the requirement that students believe it.

In the end, both sides went home disappointed. Technically, Scopes lost his case. In a quick nine minutes, the jury found him guilty as charged. However, many evolution supporters claimed a moral and intellectual victory. They claimed at the time that the trial had exposed the vapidity and illogic of Bryan's anti-evolution crusade.[14]

Of course, the Scopes trial wasn't the first public battle over the issue of evolution education during the 1920s. In addition to public squabbles like the ones between Bryan, Edward Birge, and Edwin G. Conklin, several states considered anti-evolution legislation in the first years of that decade. And, just as the legal contest in Tennessee hadn't actually been about the issue of evolution, many of the state laws proposed ranged far beyond the issue of evolution education itself. In 1922, for example, as Kentucky's state lawmakers debated a bill to ban evolution education, Frank McVey led the public opposition. McVey, president of the University of Kentucky, followed the lead of Birge and other university administrators. He appealed to the people of Kentucky, claiming that the bill would shut the state off "from all contact with the modern world," since "all the natural sciences" were based on evolution.[15]

Evolution opponent and state senator Harry F. Greene challenged McVey: "If he is not teaching evolution what is he hollering for? If the university is not teaching evolution this bill does not hit it."[16] Greene's challenge almost carried the day, politically. The bill failed by only one vote in the state House of Representatives. As the sponsor of the House version of the bill privately admitted, he had only agreed to that narrow defeat after receiving assurances that evolution would be eliminated from Kentucky's schools even without the bill.[17]

Senator Greene must have guessed that his argument would be politically potent. He might not have cared that it wasn't precisely accurate. Kentucky's bill would have done much more than simply ban the teaching of evolution. It proposed to outlaw the teaching of "Darwinism, Atheism, Agnosticism, or the Theory of Evolution."[18] What's more, a state senate amendment would have prohibited any public library in the state from owning any materials "containing such teaching that will directly or indirectly attack or assail or seek to undermine or weaken or destroy the religious beliefs and convictions of the children of Kentucky."[19] Bills such as this had much wider ambitions than merely to ban evolution education. They asserted a dominant role for traditional Protestant religiosity in America's schools. In the eyes of evolution opponents, they spelled out the proper conditions to protect students from all threats to their faith, including but certainly not limited to evolutionary theory.

Other state legislators copied Kentucky's efforts. South Carolina debated and eventually defeated a similar bill in 1922.[20] The next year, several states considered some type of anti-evolution laws. In Georgia and Texas, sweeping anti-evolution bills received favorable committee reports before meeting defeat.[21] Also in 1923, the state legislatures of West Virginia, Alabama, Iowa, and Tennessee all defeated some form of anti-evolution law.[22] In Florida, state legislators approved a nonbinding resolution that condemned any teaching challenging traditional Protestant faith.[23] Oklahoma soon passed an anti-evolution textbook law, with the strong support of the state Ku Klux Klan.[24] In the following year, 1924, California's State Board of Education issued a warning to schoolteachers to restrict their teaching of evolution.[25] The Board of Education in North Carolina voted to ban a biology book that, in its opinion, contradicted the creation story in Genesis.[26] Even the US Congress considered the issue that year. It passed an amendment prohibiting teachers in Washington, DC, from teaching any idea suggesting "disrespect of the Holy Bible."[27] Before the Scopes trial in 1925, several states considered legislation similar to that of Tennessee. In West Virginia, North Carolina, Georgia, Florida, and Texas, the bills were defeated.[28] In Texas, however, governor Miriam "Ma" Ferguson ordered publishers to edit out any mention of evolution and Darwinism from their textbooks.[29] Most famously, in Tennessee the governor signed the Butler Act into law on March 23, 1925. This act also ranged beyond the narrow issue of evolution education. It stipulated that no theory be allowed to challenge students' belief in the creation story found in Genesis.[30]

This surge of political agitation and support for anti-evolution laws fueled both evolution supporters and opponents. Supporters were shocked and saddened by the way the opposition had misrepresented evolution education. They professed consternation that anyone would hope to fight against the teaching of such a fundamental scientific idea. They might have agreed with Edwin Conklin that it was the epitome of absurdity to attempt to repeal "a law of nature with a law of Kentucky."[31] Yet, like the American Association of University Professors, a group that fervently opposed anti-evolution legislation, many evolution supporters assumed that such laws represented the rising political tide.[32]

Evolution opponents roared into the summer of 1925 confident that this surge of political opposition to the teaching of evolution could overwhelm any legal and pseudoscientific sophistries that the ACLU and the Scopes defense team could put together. Even after that summer, political organizing against evolution education continued to gain momentum. In the aftermath of the public spectacle of the Scopes trial, however, the terms of the debate hardened. Evolution supporters routinely accused their opponents of being merely ignorant, isolated hillbillies, as many reporters had described the residents of Dayton, Tennessee. Evolution opponents often took a more conspiratorial tone, denouncing evolution as a sinister attempt by scientific and educational elites to undermine the religious faith of America's schoolchildren.[33]

These debates in the aftermath of the Scopes trial sketched out positions in the conflict over evolution education that would endure throughout the rest of the twentieth century and beyond. Evolution supporters had secured influential institutional homes, including leading universities and scientific professional organizations and publications. Evolution opponents, for their part, established opposition to evolution education as a key tenet of conservative Christianity among both Protestants and Catholics. They recognized the need for independent institutions, including a separate anti-evolution scientific establishment and more congenial institutions of higher education. These weren't simple developments occurring overnight or uniformly. But the changing tone of debate over evolution education in the years immediately following the Scopes trial demonstrated that positions on the issue were hardening.

When the US Congress considered the issue in 1926, Representative Thomas Blanton of Texas insisted that teachers must not be simply "turn[ed] loose upon unsuspecting pupils" to "teach any kind of doctrine they want." Aware of the public reputation of the anti-evolution movement, Blanton added defensively, "I know the newspapers make fun of us and call us 'fundamentalists' whenever we want to inquire into what the children are being taught."[34] Evolution supporters in Congress leaped to the attack. Representative Frederick Lehlbach of New Jersey protested that anti-evolution laws forced a "most onerous and difficult duty" on teachers and ought not be hurried into. Fiorello LaGuardia, at the time representing his East Harlem district, insisted that students were "safe in the schools; they are learning to think." He blamed the "hysteria" and the "wave of intolerance" of recent fundamentalist school campaigns for enacting "ridiculous" school laws. LaGuardia hoped that Congress would not "follow the mistakes or foolish conduct of any State legislature." In short, he insisted that anti-evolution campaigners had misunderstood the nature of education. As did many evolution supporters, he believed that education must first serve to train skeptical, inquiring minds; anti-evolution laws could only block education, not reform it.[35]

Many state legislators, however, continued to push for anti-evolution laws for their states. Mississippi, Kentucky, and Louisiana considered anti-evolution bills in 1926, and the bill in Mississippi passed.[36] In 1927, when due to their biennial structure many more state legislatures met, more states considered similar laws. Legislatures in North Carolina, South Carolina, Florida, Alabama, Arkansas, Oklahoma, Missouri, West Virginia, Maine, New Hampshire, Minnesota, North Dakota, Delaware, and California all considered some sort of anti-evolution bill. After 1927, the legislative pace slackened, with only South Carolina reconsidering a bill in 1928 and only Texas in 1929.[37] However, anti-evolution activists in the Arkansas state legislature managed an end run around legislative failure. They successfully passed an anti-evolution law by referendum in 1928.[38]

As had earlier bills, this flurry of proposals reached far beyond the narrow issue of evolutionary science. Bills debated in 1927 in Alabama, Arkansas, South Carolina, and North Dakota all would have banned the teaching of evolution in public schools, including universities and colleges. But, like the Tennessee law of 1925, they also reserved a special place in schools for Protestant religiosity. The bills in Alabama, Arkansas, and South Carolina all required the prohibition of "any theory that denies the story of the Divine Creation of Man as taught in the Bible, and to teach instead that man has descended from a lower order of animals."[39] The North Dakota bill insisted on a special consideration of "the study of the Divine Creation of Man."[40]

Some of the bills debated in state legislatures in the years immediately following the Scopes trial made even more sweeping demands. West Virginia's state representatives debated a bill that would have banned "the teaching of any nefarious matter in our public schools."[41] In Florida, Representative Leo Stalnaker introduced a bill that would have banned

> the teaching as fact [of] any theory that denies the existence of God, that denies the divine creation of man, or to teach in any way atheism or infidelity, and to prohibit the use or adoption for use of any text book which [does so] . . . or that contains vulgar, obscene, or indecent matter.[42]

Both evolution opponents and supporters described all these bills as "anti-evolution." But just as the Scopes trial wasn't legally about the issue of evolution education, so these proposed laws went far beyond the narrow issue of evolution. What might "nefarious matter" include? Who would decide if materials were "vulgar, obscene, or indecent"?

This brings us to the central question: if the anti-evolution movement of the 1920s wasn't just about evolution, what was it about? In short, the battles over evolution during that decade were contests for control. This fact was recognized by leaders at the time. Clarence Darrow, for instance, as he cross-examined William Jennings Bryan at the Scopes trial, insisted that his hostile questions had "the purpose of preventing bigots and ignoramuses from controlling the education of the United States."[43]

Bryan didn't agree with Darrow on much during the summer of 1925, but he agreed that the real question at stake in the Scopes trial was not simply evolution but, as he put it, "Who Shall Control" American education. He argued that a recent US Supreme Court case had affirmed the right of parents to control the education of their own children.[44] In that case, *Pierce v. Society of Sisters*, the court ruled that a Klan-backed Oregon law prohibiting private education violated parents' rights.[45] Evolution education, Bryan argued, simply replaced the undemocratic control of the Oregon Ku Klux Klan with the undemocratic control of "a pretty little oligarchy" of mainstream scientists.[46]

Evolution education represented a clutch of ideas and trends that large portions of the American populace disliked. One anti-evolution women's group from Indianapolis articulated its reasons for opposing evolution education as nothing less than a fight "for the preservation of the public schools of America, the Bible and the faith of our fore-parents."[47] Another evolution opponent had a similarly simple yet sweeping description of the issue. "The doctrine of evolution," he explained, "is the spiritual path that leads to Sodom."[48] Banning evolution, or, for politicians, being seen as fighting to do so, indicated an opposition to a cluster of dangerous ideas and trends in schooling and culture.

Evolution supporters similarly saw the issue as central to the ultimate control of American education. As Edwin Conklin later remembered, he had fought so energetically for evolution education in the 1920s because evolution was the heart of real science. It not only formed the "central theme of biology," it symbolized science and the modern scientific mindset.[49] Not to teach evolution thoroughly and intelligently meant to bow to the "unreasoning attitude" of anti-evolutionists.[50]

Another 1920s evolution supporter, Maynard Shipley, agreed entirely. Shipley, self-designated head of the Science League of America, considered the fight against evolution education to be simply the entering wedge in a much wider "War on Modern Science." Those who opposed evolution education, in Shipley's opinion, did more than just advocate for certain types of science education. They represented "armies of ignorance," the "forces of reaction."[51] If they were allowed to ban the teaching of evolution in America's schools, these aggressive and implacable forces of "organized ignorance"[52] would "eventually dominate our public institutions."[53]

As the legislative battles of the 1920s wound down, both evolution supporters and opponents concluded that they had soundly defeated their misguided and unscientific opposition. Both sides retreated in the smug satisfaction that they had protected American education from ignorant bullies. As the perspicacious reporter Russell Owen noted at the conclusion of the Scopes trial, "Each side withdrew at the end of the struggle satisfied that it had unmasked the absurd pretensions of the other."[54]

Mainstream scientists largely retired to their research, satisfied that evolution had carried the day, except possibly among some geographically and culturally isolated outposts. In the years following the Scopes trial and 1920s legislative battles, Shipley had a difficult time convincing his allies that the battle hadn't yet been won. "Although it is continually asserted," he protested, "that 'the Dayton trial and Bryan's death ended the Fundamentalist drive' . . . this is far from being the case."[55]

Shipley was correct. Opponents of evolution education hadn't been driven from the field. In their opinion, at least, they had largely won the day. They reveled in Bryan's victory at the Scopes trial, for instance. While evolution supporters concluded that Bryan had been humiliated, evolution opponents told themselves that he had been nothing less than

> a real man, face set like a flint, eyes lighted with a radiant glow, standing four square upon his feet, with Bible in hand, ready to fight for it, ready to die for it, for he had tested it, proved it, believed with all his heart that it was worth defending to the death.[56]

Of course, Bryan died just a few days after the conclusion of the trial, and many of his allies agreed that he had died a hero. After all, as they surveyed the field of evolution education at the end of the 1920s, it looked to them as if they had defeated the bugbear once and for all.

They often had a good deal of evidence to support that conclusion. The controversy had convinced many cautious educators and educational publishers to avoid the subject entirely, effectively banning the topic of evolution from many of the nation's schools. Although only five states—Oklahoma, Florida, Tennessee, Mississippi, and Arkansas—had passed laws or resolutions limiting evolution education, anti-evolution activism had frightened many textbook publishers into changing their curricular offerings. One representative of Ginn and Company, for instance, had written to Bryan near the start of the 1920s controversies. He complained that his company had had several of its titles banned on suspicion of promoting evolution. He hoped Bryan would affirm that teaching evolutionary theory was proper; it was only the teaching of the theory as scientific fact that was provocative. Bryan demurred. Although he did indeed admit that older students ought to learn about evolutionary theory, he didn't agree to lend his reputation to any textbook series.[57] Even when evolution opponents hadn't succeeded in passing state laws banning evolution from schools, they had been able to frighten textbook publishers and local educators into treating the issue very tentatively if at all.

Over the course of the next several decades, each side in this lasting tension could be justified in believing that the contests of the 1920s had settled the issue in their favor. As the next chapter will describe, mainstream scientists and other evolution supporters deepened their reliance on evolution as a basic building block of modern science. They also consolidated their control of key institutions such as mainstream higher education and professional organizations. Meanwhile, evolution opponents could usually look at their local schools and conclude comfortably that evolution was no longer a problem. It was not until the 1960s that a series of events jarred both sides from their complacent reveries.

CHAPTER 3

The Dog That Didn't Bark

Historians have a favorite Sherlock Holmes story. In the well-known tale "The Adventure of Silver Blaze," Arthur Conan Doyle has his famous detective solve an apparent horse theft with a key clue: a guard dog that did not bark. We won't give away the details of the mystery, but for historians, the lesson is clear. Often, the most important piece of information is the clue that isn't there.

In the case of the continuing struggle over evolution education in the twentieth century, the dog that didn't bark was the long period between roughly 1930 and 1960 in which the issue disappeared from the nation's headlines. However, people hadn't stopped caring about it. Nor had evolution opponents—despite the fond wishes of evolution supporters—agreed that they had lost the fight in the public controversies of the 1920s. Rather, the issue did not surface in public controversies largely because both sides concluded that they had won. Both sides had plenty of evidence to support their belief. Evolution supporters noted breathtaking scientific advances, including a new consensus among mainstream scientists about the ways natural selection worked. Opponents took solace from the fact that their local schools and textbooks didn't teach much evolutionary theory.

It was only in the early 1960s that this deluded truce was breached. In the Sputnik-fueled drive for improved science education, the federal government backed a series of textbooks that challenged the timid policies of commercial publishers. These textbooks, the Biological Sciences Curriculum Study, made evolution one of their central themes. As the books and their reputation spread, a new wave of opposition to evolution education emerged. Also in the early 1960s, two US Supreme Court decisions forced many evolution opponents to reconsider their relationship with public schooling. Although the decisions didn't deal specifically with the issue of evolution education, the court's rulings in the cases of prayer and Bible reading in the schools had an enormous impact among conservative Christians, often the same activists who led opposition to evolution education.

In the 1930s, mainstream scientists made advances in their understanding of genetics that helped solve a long-standing dispute over the mechanism of evolution. Earlier scientists, including Charles Darwin himself, had operated on the premise that offspring blended the characteristics of their parents. This presented a few problems for the theory of natural selection, especially the likelihood that any advantageous mutation would quickly become swamped by interbreeding with parents who didn't share that mutation. It seemed impossible that any new mutation would persist.[1] For this and other reasons, as we have seen, mainstream scientists in the 1920s often spoke of the end of Darwinism.

The answer eventually came by a circuitous route from some pea plants. In the nineteenth century, Gregor Mendel's studies of these plants hadn't attracted immediate attention from the scientific community. By the turn of the twentieth century, however, a new generation of scientists, Hugo de Vries in particular, recognized that Mendelian genetics could solve some of the problems with the mechanism of natural selection.[2] By the 1930s, scientists such as Ronald A. Fisher, J. B. S. Haldane, and Sewall Wright forged a new synthesis of Darwin's idea of natural selection with Mendelian genetics.[3]

Throughout the 1930s, mainstream scientists such as George Gaylord Simpson and G. Ledyard Stebbins worked, in the words of historian Edward Larson, to "institutionalize their approach to the study of evolution within the mainstream scientific community."[4] In the decades that followed, leading scientists Theodosius Dobzhansky, Ernst Mayr, and E. B. (Henry) Ford were among those who cemented the supremacy of this explanation of evolutionary mechanisms and its unchallenged dominance in professional scientific institutions and publications.[5]

By the late 1950s, mainstream scientists had become confident in what they called the neo-Darwinian or modern evolutionary synthesis. Advantageous mutations could be passed along through natural selection, with recessive genetic traits surfacing relatively quickly when they bestowed an evolutionary advantage.[6] The details of evolution still offered plenty of room for scientific research, debate, and disagreement, but the basic story of life-forms developing from earlier forms based on advantageous mutations had become established as a fundamental truth of biology. No mainstream scientist still argued that life on earth, even human life, had been created by fiat.

By the 1960s, mainstream scientists were more confident than ever that evolution explained the origins of life. In the words of Simpson, whose paleontological research had contributed a great deal toward establishing the modern evolutionary synthesis, "Organic evolution is one of the basic facts and characteristics of the objective world."[7] However, just as Edwin Conklin had realized a generation before, Simpson recognized that this scientific consensus didn't carry much weight outside its own professional circles. Following in Conklin's footsteps, he hoped to carry his evolutionary message to that wider audience. For decades, beginning in the 1940s, Simpson tried to cure the general public of its false belief in a world not created by evolution.[8] He recognized that Darwin's theory, even in its modern synthesis, had reached "alas!—only a minority of us."[9]

Part of the reason for this broad gulf between mainstream scientific opinion and that of the majority of the populace was the fact that the public controversies over evolution education in the 1920s had largely scared educators and publishers away from the issue. As John Scopes noted glumly in his memoirs, even forty years after the Scopes trial he would probably still be punished if he tried to teach evolution. Teachers in Dayton, Tennessee, still had to sign pledges that they would not teach the subject.[10] Even "in the Space Age," the older Scopes lamented, evolution opponents kept the theory far from their children's classrooms.[11] Simpson agreed. Despite it being the 1960s, he complained, "innumerable students still leave high school without ever having heard of evolution, or with having heard of it only in such a way as to leave them unimpressed or antagonistic."[12]

Generalizations about what has been taught in America's classrooms are notoriously inexact. The localized nature of school decisions in the United States makes it difficult for historians to assert with any great confidence what was and what was not taught in classrooms in the past. Individual school districts can and do impose policies that differ from those a few towns over. And individual teachers have a great deal of control in what they actually teach. In short, it is very hard for historians to get behind those classroom doors.[13] If they worried that their choices might be controversial, teachers wouldn't be likely to publicize their decisions. There may have been a great deal more evolutionary education going on in the 1930s, '40s, and '50s than is evident in the historical record. However, despite those important caveats, it seems that Simpson's and Scopes's impressions were correct. Just as mainstream scientists achieved successes in reconciling Mendelian genetics with Darwinian selection, and in establishing that evolutionary synthesis as a basic building block of mainstream science, so evolution opponents had largely been able to keep those ideas out of America's secondary science classrooms.

Historians of religion have argued that after the public controversies over evolution education in the 1920s, many conservative Protestant evangelicals retreated in humiliation from public activism to build their own institutions.[14] Some did. As we will explore below, many more evolution opponents simply felt that they no longer needed to protest against evolutionary education. Often, their local schools gave them good reason for this complacency. Most historians have agreed that one of the most serious casualties of the Scopes trial was the treatment of evolution in school textbooks. After Scopes, the story usually goes, mainstream textbooks scrubbed out any mention of evolution. There is much truth to this telling, but as historian Adam Shapiro has recently argued, the actual process was much more convoluted. Textbook authors often fought—and often successfully—to keep basic evolutionary content the same in post-Scopes revisions. But sales agents with an eye on the bottom line pushed hard to purge their product of any poisonous content that might attract controversy and discourage sales.[15] In the end, textbooks often reflected this cacophony of interests. Many successful science textbooks changed wording and cleaned indexes to eliminate controversial terms such as *evolution* and *Darwin*. But many of those same books kept most of their evolutionary content. Busy book watchdogs seemed satisfied to check indexes and scan text superficially. In one case, Shapiro found, a post-Scopes textbook that contained very little evolutionary content became controversial only because its editors had forgotten to eliminate the term *evolution* from an attached glossary.[16] Parents worried about evolution might with justification feel confident that their children's textbooks had been purged of evolution, even if that purge had actually been superficial.

During this lull in public controversy over evolution education, those who had retreated to new independent institutions helped create a new anti-evolution infrastructure. Just as mainstream scientists in the 1930s, '40s, and '50s established the dominance of the modern evolutionary synthesis at leading research universities and in scientific professional organizations, so these new anti-evolution institutions established a new consensus among evolution opponents during that same period. Thus, by the time that public controversies over evolution education broke out again in the 1960s, evolution's opponents had changed just as much as its supporters had.

One important type of new institution was the Protestant fundamentalist college or seminary. The mission of these schools wasn't limited to the issue of evolution, but their growth and influence had an enormous impact on the development of anti-evolution sentiment during the decades following the 1920s controversies. Many of them trained a generation of students in an explicitly anti-evolution worldview. They provided homes for intellectuals who opposed evolution and evolutionary education. They trained pastors in a theology that insisted on steadfast opposition to evolution education. And they allowed students and families to continue their education and their pursuit of professional careers without accepting the scientific consensus about evolution that had come to dominate mainstream higher education.

An example of such a school was Bryan College in Dayton, Tennessee. As its name and location suggest, its founders wanted a memorial to William Jennings Bryan and his heroic stand at the Scopes trial. The details of the new college's theology and location caused some debate among leading 1920s evolution opponents.[17] Nevertheless, all its early supporters agreed with its first president, George Guille, who insisted it would become "internationally known for its belief in the Bible as the inspired Word of God and for its devotion to the Lord Jesus Christ."[18] Bryan College's founders inscribed those beliefs in a creed, to which students, faculty, administration, and staff had to agree. One tenet was that the new school would teach only the creation story as recorded in the book of Genesis.[19]

Other new anti-evolution colleges and universities in the 1920s insisted on similar creeds. One of the most influential, Bob Jones University, opened in 1926 with one that could "never be amended, modified, altered, or changed." Among its provisions were the notions that the school was dedicated to "combating all atheistic, agnostic, pagan and so-called scientific adulterations of the Gospel" and to teaching "the creation of man by the direct act of God."[20]

These new colleges and universities were joined by some older schools that signed on to an explicitly anti-evolution theology. Gordon College in the Boston area, the Bible Institute of Los Angeles (later Biola University), Wheaton College in suburban Chicago, and the Moody Bible Institute of Chicago were among those contributing to the new network of institutions of higher learning.[21] The scope and influence of this new network are hard to judge. Just as mainstream scientists disagreed about the details of evolution, so anti-evolution institutions have had sharp disagreements about theology, science, and political strategy. But the network of anti-evolution colleges and universities has certainly grown into a significant alternative presence in the field of American higher education. One indicator of their success has been the large numbers of adults with college degrees who have not accepted the mainstream scientific consensus about evolution. Polls have shown that just over a quarter of college graduates think humans were created by God at some point in the last ten thousand years.[22] Of course, students at secular colleges may also refuse to accept evolutionary theory, but the network of colleges and universities that have dedicated themselves to opposing evolutionary theory since the controversies of the 1920s have made it easier for American adults to be educated without being encouraged to accept the evolutionary ideas of mainstream scientists.

Beyond higher education, another outgrowth of the 1920s evolution education controversies was a new set of alternative scientific organizations. George McCready Price, the self-described geologist whom Clarence Darrow mercilessly ridiculed in absentia at the Scopes trial, briefly continued in that decade to contend with mainstream scientists to delegitimize evolutionary theory. In a debate in London just after the trial, for example, Price mocked "the specialist in some corner of science" who continued to explore the field of evolution. Such pseudoscientists, in his opinion, were "living in a fool's paradise." Due to Price's research, the foolish evolutionist would "wake up some fine morning and find that he needs an introduction to the modern scientific world."[23]

To his dismay, the London audience heckled Price viciously. Their furious shouting and relentless stamping ultimately kept him from completing his rebuttal. He never debated in public again.[24] But that didn't mean he retired from his anti-evolution activism. Price continued writing and publishing prolifically. He also helped found a new organization, the Deluge Geology Society (DGS). He realized after the public tumult of the 1920s that mainstream science had turned irresistibly in favor of evolution. In order to continue his fight, he would need his own organization.[25]

The long-term impact of the DGS was enormous. In addition to attracting immediate support from a small but energetic group of dedicated evolution opponents, it provided inspiration and an institutional home for the next generation of anti-evolution activists. Henry Morris, a civil engineer by training and a passionate opponent of evolution education, joined the DGS as he was beginning his career in anti-evolution activism. He adopted co-founder Price's fairly radical positions, including the notion of a young earth and of a literal worldwide flood. Those notions had been uncommon among anti-evolutionists of the 1920s.[26] However, Price's involvement in providing an alternative scientific institution helped establish "flood geology" as the new orthodoxy among evolution opponents.

In 1961, just as George Gaylord Simpson confidently told audiences that evolution had been established as one of "the basic facts and characteristics of the objective world,"[27] just as the Biological Sciences Curriculum Study produced its first set of evolution-heavy school textbooks funded by $10 million in taxpayer money,[28] Henry Morris and John Whitcomb Jr. published their landmark work, *The Genesis Flood.* This text brought the ideas of flood geology to a new generation of evolution opponents, a generation that often felt as if its hard-won victories over evolution education had come under renewed attack. Just as evolution proponents had created a new consensus about the likely process of natural selection, so evolution opponents achieved a new religious consensus. For significant portions of the conservative Christian population, sincere religious belief came to require belief in creation by divine fiat only a few thousand years in the past.

In the decades that preceded those 1960s controversies, however, many evolution opponents concluded that they had won their battles to rid their schools of evolutionary theory. In addition to the five states—Oklahoma, Florida, Tennessee, Mississippi, and Arkansas—that had passed anti-evolution laws or resolutions, several other state boards of education demanded that evolutionary material be deleted from their textbooks, including California, North Carolina, Texas, and Louisiana.[29]

Publishers didn't hesitate to accommodate their demands. Such states not only represented enormous textbook markets, they also served as a market bellwether for other school textbook purchasers. The 1926 edition of George William Hunter's *Civic Biology*—the older edition of which became the center of the Scopes trial—eliminated the word *evolution* from much of the text and index.[30] For example, a paragraph bearing the heading "Evolution of Man" in the 1914 edition was now preceded instead by "Development of Man."[31] Similarly, an index entry for "Darwin and Natural Selection" was revised as "Darwin, Charles, on heredity and variation."[32] Henry Holt made similar changes to Truman Moon's *Biology for Beginners*. A special edition for use in Texas schools cut three chapters that discussed the "Descent of Man."[33] Those chapters were left in the edition sent to other states, but even there, Moon made other changes. The 1921 preface declared that biology was "based on the fundamental idea of evolution."[34] In 1926, the editors changed this to "based on the fundamental idea of development."[35]

The book that one study has called "the most widely used text in the years following the Scopes trial" did not need as many alterations.[36] For the 1929 edition of Smallwood, Reveley, and Bailey's *New General Biology*, the book's already scant treatment of evolution was shortened, and the term was removed from its index. In fact, that reticence to pay much attention to the topic in the first place was one of the main reasons for the book's popularity. In the words of Charles E. Bacon, its publisher's representative, the Texas State Text Book Commission approved *New General Biology* because it had been "'tactfully' written and required no alterations."[37]

Such market-conscious modifications continued throughout the 1930s, '40s, and '50s. In most cases, textbook authors and publishers changed content to remove controversial terms such as *evolution* and *Darwin*. They preferred to treat the subject briefly and blandly. For instance, of fifteen textbooks commonly adopted by state boards of education in the 1930s, fourteen briefly discussed the idea of human evolution. But they generally used euphemisms to describe it, such as "the early history of man upon the earth."[38] In the 1940s, popular textbooks increased their inclusion of evolution somewhat. Of fifteen such books, only three included the actual word *evolution* in the text, index, or glossary.[39]

In the 1950s, coverage decreased in some cases. For example, Ella T. Smith's 1938 edition of *Exploring Biology* informed readers, "Evolution is a fact. Plants and animals do change and have been changing."[40] In this edition, she also told readers that for humans, too, "the fossil evidence is conclusive that man himself did not appear suddenly on the earth in his present form, but has gradually developed from a much more primitive species."[41] In the 1954 edition of this book, however, Smith gave a more nuanced description. She noted that evidence "leads scientists to the conclusion that the plants and animals of today are the changed descendants of the plants and animals of the past." There was another word to describe that change: "That word is *evolution*."[42] But Smith tended to use the word *change* instead. When she described "the modern point of view," for instance, she gave a bland description of evolution: "Biologists agree today that plants and animals have changed in the past, and continue to change."[43] By the 1959 edition, numerous references to evolution in the index had been reduced to one line. In the text, Smith informed readers only that "the history of living things is a long one. Much of it is still unknown."[44] Moreover, a ten-page section on the history of evolutionary theory that was in the 1954 edition was eliminated entirely. In its place, Smith offered a brief suggestion that students write a report about evolutionary theorists such as Darwin, Alfred Russel Wallace, Lamarck, or Hugo de Vries.[45]

Despite this "tactful" silence on the part of educational publishers and authors, some teachers and some school districts still may have provided a more thorough evolution education. It is impossible to judge with certainty what was taught merely from what appeared in leading textbooks. Nevertheless, the awkward, tentative treatment of evolution in major textbooks between 1926 and 1960 demonstrates publishers' sensitivity to market demands. Their bland, brief inclusion of evolutionary ideas allowed them to tell state and district boards of education that their books did not promote evolution.

A few intrepid academics attempted to gauge the depth of evolution education during these decades of relative quiet. None of their studies was quite thorough enough, to be sure, but taken together they convey an impression of deafening silence about evolution in a majority of America's classrooms. One ambitious nationwide survey in 1942 considered 3,186 responses given by secondary-school biology teachers to a lengthy questionnaire.[46] The survey authors concluded that evolution was taught in "*notably less than half* of the high schools of the United States."[47] And of those schools in which evolution was being taught, the authors concluded that usually it was "frequently diluted beyond recognition," either by pairing it with the teaching of special creation, or by separating human origins from the idea of evolution.[48]

Some of the reasons teachers gave for avoiding the topic are illuminating. Approximately 8 percent (73 of 916 respondents to this question) incorrectly believed that their state had banned evolution education. Others cited instructions from their local board of education or school-building administration, their perception of the will of the majority of their community, or their personal beliefs. The wide variety of reasons given beyond these categories doesn't tell us much in terms of raw numbers, but it demonstrates that the intensity of feeling about evolution education was a nationwide phenomenon. One respondent from Upstate New York claimed to teach evolution, noting, "I've had fights but haven't lost yet." A teacher from North Carolina stated that evolution education was a "taboo subject to most people." Other teachers echoed the response of one from rural Nebraska, who said he or she avoided evolution education mainly due to "lack of time." And one California teacher added, "Controversial subjects are dynamite to teachers."[49] Given the scope of this survey, it is not surprising that teachers revealed a wide variety of experiences with evolution education.

Other studies were less ambitious. In 1950, a Newark, New Jersey, high-school science teacher and a Rutgers University–Newark faculty member teamed up to investigate what they called "the meager, evasive, and equivocal discussion" of evolution in high-school biology textbooks.[50] Their survey of Newark teachers and textbooks confirmed their suspicions. The biology textbooks available to science teachers in that city often left out the words *Darwin* and *evolution* entirely.[51] One informed readers that plants and animals "were not created at one time." As the authors of the study noted, the lingering insistence on using the verb *create* "may make for sound doctrinaire theology, [but] it is certainly not biology."[52] Other available textbooks, notably Smith's *Exploring Biology* (1943 edition), included a more thorough treatment of the idea of evolution. Even so, roughly one-third of the survey respondents (eight of twenty-nine) claimed to make no mention of evolution in their classroom teaching. And of the teachers who did claim to teach evolution, they taught on average merely a fraction of the days recommended for the subject by the city's standard science curriculum—six days instead of ten.[53]

Neither the Newark nor the nationwide study could make comprehensive claims about the amount of evolutionary education going on across the country. But they do suggest that teachers and textbooks tended to downplay evolution education.

One additional piece of evidence suggesting the lack of evolution education during this period was "the dog that didn't bark" alluded to at the beginning of this chapter. Between 1930 and the 1960s, no public controversies over evolution education gained national attention. This wasn't because evolution opponents had retreated from any public battles over schooling. To the contrary, when issues such as weekday religious education attracted their attention, conservative Christians became very active in educational politics in the 1930s, '40s, and '50s.[54]

Scattered evidence of local controversies also suggests that evolution opponents continued to keep a wary eye on local and state education policy. In 1937, the state superintendent of schools in Indiana responded to public pressure by insisting, "The theory of evolution should not be advocated, and frankly I doubt if it is advocated in any school in Indiana."[55] That same year, public-school officials in Philadelphia responded to anti-evolution pressure by denying that human evolution was being taught in that city's public schools.[56] Evolution opponents certainly maintained a reputation for contentiousness that made these school officials squirm.

Although we can't be certain, evolution opponents likely did not challenge the teaching in their local schools more often because they felt that there was not much evolution education to challenge. Perhaps paleontologist George Gaylord Simpson hit the nail on the head when he lamented in 1961, "Although almost everyone has heard of [evolution], most Americans have only the scantest and most distorted idea of its real nature and significance."[57]

In 1957, the successful launch of the Soviet Union's Sputnik satellite spurred a renewed interest in American science education. The federal government poured millions of dollars into the development of a new series of science textbooks, the Biological Sciences Curriculum Study (BSCS).[58] Due to such funding, the editors of the series were free to ignore the market considerations that had so intimidated commercial publishers.[59] As a result, they made evolution one of the new textbooks' nine central themes.[60] By the late 1960s, nearly half of American high schools were using BSCS materials to some extent.[61] Just as important, commercial publishers rushed to update and deepen their own treatment of evolution to keep up with that series.[62]

The BSCS textbooks forced evolution supporters and opponents to notice each other again. After decades of relative quiet, evolution opponents could not help but notice the way mainstream science had hardened its position in support of the theory of evolution through natural selection. In the 1920s, anti-evolution activists confidently told audiences that mainstream scientists would come around shortly. Opponents assumed they soon would recognize that their dabblings in evolutionary theory were not leading them anywhere.[63] But by the 1960s, anti-evolutionists could no longer maintain such fond hopes. The publication of the BSCS textbooks reinforced their fears that mainstream science had been permanently and thoroughly taken over by evolutionary theory.

However, the BSCS textbooks weren't the only change that affected the reemergence of public controversies over evolution education. Other events had a significant impact on how evolution opponents felt about public education, in ways that most histories of evolution education have not adequately recognized. Two of the most important were a pair of US Supreme Court decisions. In the 1962 *Engel v. Vitale* case, the court decided that a nonsectarian prayer composed by the New York State Board of Regents could not be used in public schools.[64] In 1963, in *School District of Abington Township v. Schempp*, the court made an even more profound statement about religion in public schools. It ruled that Bible reading and use of the Lord's Prayer, even when students were excused with parental permission, unacceptably breached the wall of separation between church and state. Before the 1963 decision, only ten states had passed laws against Bible reading in public schools.[65] The decision reflected a new understanding of the proper constitutional relationship between religion and government. As justice Tom C. Clark wrote in his majority opinion, any legislation must have a "secular legislative purpose and a primary effect that neither advances nor inhibits religion."[66] This was a higher hurdle for evolution opponents, and one that led to their increasing claims of scientific merit for the curricula they proposed.

Not all evolution opponents were conservative evangelical Protestants. Nor were all evangelicals opposed to the teaching of evolutionary theory. Nevertheless, opposition to evolution education, along with support for Bible reading and vocal prayer in the classroom, remained central educational policy positions for many American evangelicals. The Supreme Court decisions of 1962 and 1963 left many evolution opponents in agreement with the editors of the leading evangelical magazine *Christianity Today*. In the aftermath of the *Schempp* and *Engel* decisions, those editors quickly moved from a celebration of prayer in the public schools as the inviolable demand of America's "devout masses"[67] to a defensive exhortation that evangelicals formed only a "believing remnant" in American culture.[68] Evolution opponents often agreed that they could no longer simply demand that evolution be prohibited from public schools. No longer could they insist, as they had in the public controversies of the 1920s, that schools ban evolutionary theory.

In order to understand the public controversies over evolution education that surged back onto the nation's headlines in the 1960s, it is vital to understand the changing attitudes that caused this resurgence. Just as the new series of BSCS textbooks convinced many evolution opponents that mainstream science and science education had gone irretrievably to the side of evolutionary theory, so the Supreme Court decisions of the early 1960s forced them to acknowledge that their potential to influence public education had changed dramatically. Changing ideas about science, theology, and culture would play a decisive role in the new round of controversy over evolution education.

Yet such ideas didn't spring to life the instant the BSCS textbooks entered the nation's classrooms. Rather, they were a long, slow, quiet development that neither evolution opponents nor supporters ordinarily recognized at the time. They were the dogs that didn't bark. During the 1930s, '40s, and '50s, both sides of the evolution issue consolidated their hold over their respective institutions. Meanwhile, due to educators' and publishers' worry about provoking controversy, the actual teaching of evolutionary theory remained hit-or-miss in the public schools. As mainstream scientists worked out the kinks in the theory of evolution through natural selection, so anti-evolution scientists embraced a theory of origins that insisted on a young earth and a recent, catastrophic worldwide flood. By the time the BSCS textbooks brought public attention back to the issue of evolution education, both sides had established themselves as durable, internally coherent belief systems with extensive political support and strong institutional structures. But the new round of public controversies would put both sides to the test.

CHAPTER 4

A New Minority

Duane Gish was not happy. It was 1984, and he had accepted an invitation to attend a meeting hosted by his scientific archenemies, the American Association for the Advancement of Science. When he arrived, he found that he was one of only two creationists in attendance. To Gish it felt like a trap, and he grumbled that he would "proceed to take one of the two seats on the back of the bus reserved for the creationists in this meeting."[1]

In his complaint, he articulated a new self-perception common among evolution opponents since the 1960s. Renewed decades of combat against evolution education had transformed their positions. In the 1920s, as we have seen, evolution opponents sought to use their majority influence to ban evolution from public schools and colleges. In the 1930s, '40s, and '50s, they largely concluded that at least their local schools were safe from the threat of evolution education. In the 1960s, however, that confidence was shattered by new textbooks and new rules about prayer and Bibles in public schools. By the 1980s, like Gish, many evolution opponents took the position of a beleaguered minority, shunted to the back of the cultural bus. Their earlier demands changed to demands for fairness, for equal rights, and for consideration of the idea of special creation alongside evolution in the nation's schools. By the end of the twentieth century and the start of the twenty-first, in other words, evolution opponents had adopted the rhetoric of an earlier generation of evolution supporters.

In 1995, Gish appropriated the 1925 Scopes trial line of defense attorney Dudley Field Malone. The defenders of evolution back then, Gish correctly pointed out, had insisted "that it was bigotry to teach only one theory of origins."[2] Gish was not the only latter-day evolution opponent to accuse evolution supporters of bigotry and closed-mindedness. In 2004, a controversy erupted in Dover, Pennsylvania, over the teaching of a new type of anti-evolutionism. "Intelligent design" proponents promised to keep out the religion and teach the purely scientific criticisms of evolution. The Dover school board passed a resolution that year that would have asked students to "keep an open mind" about the possible meanings of evolution and the means by which life may have originated. The board suggested that students consider the idea of intelligent design.[3]

Since the 1960s, when this latest round of evolution controversies began, evolution opponents have portrayed themselves as the put-upon minority in the debate. From the fights over "creation science" through the twenty-first-century debates over intelligent design, they took the positions occupied by evolution supporters in the 1920s. Fighting to include creation-friendly ideas about human origins became a fight for academic freedom, open-mindedness, and fairness.

By the 1960s, evolution opponents confronted a scientific establishment that had become newly aggressive in its drive to spread evolution education. New professional organizations for secondary-school science teachers took up the cause. In the high-profile court cases that resulted, these organizations successfully refuted all claims that creation science or intelligent design represented reputable alternative scientific theories. Nevertheless, evolution opponents in the last few decades of the twentieth century managed to retain significant political support.

As we have seen, one of the causes of this new round of controversy was the new Biological Sciences Curriculum Study textbooks. The BSCS series got its first taste of the dispute as soon as its directors test-marketed some of the books in 1961, and that dispute didn't focus only on evolution. In Dade County, Florida, school officials asked the series editors to remove a chapter about human reproduction. The officials had concluded that parents would be offended by the explicit diagrams of male and female reproductive organs. BSCS leaders declined. They had resolved to produce a new type of textbook, one that would not bow to political or cultural pressure. Furthermore, the BSCS leaders insisted that if Dade County school officials cut out the pages, they would be guilty of a breach of contract. The school leaders proved equally stubborn. They called in teachers over Christmas break to laboriously black out the diagrams by hand.[4]

But despite the media attention lavished on this standoff, the real focus of controversy over the new BSCS textbooks came from their treatment of evolution. In Arizona, the test-market editions of the books provoked a series of public protests throughout 1962, 1963, and 1964.[5] In New Mexico, the state board of education insisted that the books be stamped on the inside front cover with a warning that evolution was not a fact, only a theory.[6] There were also "scattered complaints" about the early edition of the books in Indiana, Alabama, Minnesota, and Kentucky.[7]

The biggest dispute over the BSCS textbooks came in Texas. In 1964, the State Textbook Committee's debate about the books led to months of ferocious public battles over evolution and the nature of the new books.[8] The outcry attracted so much attention that the meetings of the textbook commission were broadcast on national television.[9]

The nation had changed since 1927, however. The State Textbook Committee ruled that the books were to be adopted for possible use in Texas schools. Similarly, the books were adopted in Arizona, Alabama, Minnesota, Kentucky, and New Mexico, in spite of vociferous and angry protests against them. Even those who might not have felt particularly strongly about evolution education often recoiled at the aggressive activism of evolution opponents in those states. They might have agreed with superintendent Howard Seymour, of Phoenix (Arizona) Public Schools, who insisted that the books would be adopted and the schools would always welcome "a spirit of inquiry and a search for truth."[10]

By the 1960s, evolution opponents could no longer summon the political influence to simply get the books banned. Those who tried a different approach, however, enjoyed significant success. In California, parent activist Nell Segraves had pushed since 1963 for public schools to include both creation and evolution.[11] With the support of State Superintendent of Public Instruction Max Rafferty, Segraves argued that such dual instruction was the only way to respect the rights of creationist children.[12] Rafferty declared that these students ought to be protected, based on the 1964 Civil Rights Act.[13] He insisted that evolution-only curricula ignored the minority rights of creationist students.

The push for equal time for the teaching of creationism alongside any evolution education generated a great deal of political support nationwide. Between 1963 and 1978, eleven states—Arizona, Michigan, Colorado, Wisconsin, Georgia, Washington, Tennessee, Indiana, Kentucky, Ohio, and Texas—considered a total of twenty-six evolution-education bills.[14] Twenty-two of these bills specified equal time for both evolution and creationism.[15] Two became law: Tennessee in 1973 and Kentucky in 1976. In Georgia, a successful resolution in 1974 urged the state board of education "to establish policy requiring the teaching of divine creation in the public schools of Georgia."[16] In addition, evolution opponents in three states—Arizona, Colorado, and Washington—attempted unsuccessfully to put the evolution issue directly before voters by referendum, initiative, or amendment.[17]

The careers of these state bills illustrated the changing nature of the controversy. The 1964 initiative in Arizona would have banned evolution education entirely.[18] Activists failed to secure enough signatures to get this measure on the ballot. In Tennessee, the new two-model law had appealed to lawmakers as a fair inclusion of two different versions of science.[19] In spite of such demonstrated political support, the Tennessee Supreme Court concurred with lower courts and ruled the two-model approach unconstitutional.[20] What its state's politicians had seen as a fair way to assuage continued strong opposition to evolution while still teaching good science, its state's judges called an imposition of religion in public education. The Kentucky law, passed in 1976, did not ban evolution. It did not even mandate equal time for both evolution and creation. Rather, it only specified the legal right of Kentucky teachers to teach biblical creation if they chose.[21]

These legal cases demonstrated the new power of evolution supporters. Politicians and activists could no longer muster the backing to ban evolution from public schools. At most, evolution opponents succeeded in asserting the right of teachers to discuss other ideas alongside evolution. This newly dominant position for evolution supporters resulted in part from a new confidence among mainstream scientists. By 1959, in the words of historian Edward Larson, evolution had become "dogma" among this group.[22] Even more than they had during earlier evolution controversies, in Larson's opinion, many evolution supporters evinced a missionary zeal to propagate evolution education.[23] They came to utterly dominate mainstream institutions of higher education and scientific professional organizations.

One of the leading popularizers of evolutionary theory, paleontologist George Gaylord Simpson, articulated an analogy common among evolution supporters. To allow people to continue to disbelieve in evolution, Simpson argued in 1960, was like allowing adults to continue to believe in Santa Claus. Although such beliefs might make them happy, he maintained, "adults should prefer to live in a world of reality and reason."[24]

Other evolution supporters insisted with leading geneticist and biologist Theodosius Dobzhansky that education without a thorough grounding in the ideas of evolution was worse than useless. As Dobzhansky quipped to a meeting of the National Association of Biology Teachers in 1973, "Nothing in biology makes sense except in the light of evolution."[25] Similarly, a group of scientific luminaries including Simpson, Isaac Asimov, Linus Pauling, and Ernst Mayr signed a "Statement Affirming Evolution as a Principle of Science" in 1977. In an attempt to demonstrate the depth of scientific support for evolutionary theory, the signatories affirmed,

> There are no alternative theories to the principle of evolution. . . . Moreover, the principle is so important for an understanding of the world we live in . . . that the public in general, including students taking biology in school, should be made aware of it, and of the fact that it is firmly established in the view of the modern scientific community.[26]

This kind of statement often seemed redundant to many evolution supporters. Evolutionary theory had come to represent such a central idea in the science of life that many mainstream scientists assumed there could be no dissent. As a result, these scientists failed to prepare for the coming cultural storm. Indeed, historian of science Judith Grabiner and high-school math teacher Peter Miller criticized mainstream scientists for their complacency. If only more scientists had been aware of the continuing strength of anti-evolution sentiment, Grabiner and Miller argued, perhaps some of the vicious dissention might have been prevented.[27]

Just as evolution opponents had established institutions to support their activism, so evolution supporters scrambled to spread their certainty about the need for thorough and up-to-date evolution education beyond the circle of elite scientists. Groups such as the American Association for the Advancement of Science, the American Civil Liberties Union, the National Education Association, and the American Association of University Professors had long fought against restrictions on all kinds of teaching. In the 1960s, two similar professional organizations devoted more attention to promoting and defending evolution education. The National Science Teachers Association (NSTA) had been formed in 1944, but did not become active in evolution controversies until the 1960s.[28] The NSTA involved itself in high-profile legal cases and often contributed briefs meant to inform judges of the firm consensus about evolution among scientists and science teachers.[29] Similarly, the National Association of Biology Teachers (NABT) had been established in 1938. But it was not until 1967 that it set up an important committee on academic freedom. Then in 1971, the NABT created a new Fund for Freedom in Science Teaching.[30] These organizational changes allowed the NABT to take the lead in legal controversies over evolution education, such as its early challenge to the 1973 two-model law in Tennessee.[31]

Professional organizations demonstrated their new confidence by seeking out ways to overthrow lingering legal opposition to evolution education. In 1965, the Arkansas Education Association sought to challenge state law. Its new biology textbooks, following the lead of the Biological Sciences Curriculum Study series, gave much more consistent emphasis to the idea of evolution. In order to teach from them, Arkansas teachers would have had to break the state's 1928 anti-evolution law, still on the books but not enforced.[32] In the words of Susan Epperson, the first-year biology teacher who agreed to serve as the plaintiff in the case, the old law was a shame, a continuous "cloud over [teachers'] heads."[33] She and other Arkansas teachers were no longer willing to continue the charade of teaching evolution quietly and illegally.

When the case reached the US Supreme Court in 1968, evolution supporters submitted briefs summarizing their new confidence in the status and importance of evolution education. The NSTA's brief appended a statement signed by 179 leading biologists, which assured the court that "scientists and other reasonable persons" no longer questioned the validity of evolution.[34] By prohibiting instruction in such a "fundamental scientific principle," the brief insisted, the law squeezed out true education and replaced it with a "deadening dogma."[35] According to mainstream scientists, evolution had become not only a fundamental truth of science but a basic building block of knowledge that must be the foundation of any legitimate education.

The Supreme Court agreed. In his majority opinion, justice Abe Fortas noted that any attempt to prohibit such a fundamental idea seemed "quixotic" and uncomfortable to "the modern mind." Any notion that such fundamental beliefs of mainstream science ought to be restricted for religious reasons had no place, Fortas asserted, in up-to-date education.[36]

Thoughtful evolution opponents noted the trend of these court cases and state laws. In 1978, Yale law student Wendell Bird articulated a new legal basis for evolution opposition in the *Yale Law Review*.[37] The first leg of this new legal strategy noted the trend of the US Supreme Court's post–World War II reasoning. Since the 1940s, the court had repeatedly struck down school laws that did not have a primarily secular purpose. Therefore, Bird reasoned, creationism must be included if its purpose was primarily to improve science education. "Creation science" presented itself as having that purpose. It relied not on biblical arguments but rather only secular arguments to teach students the purely scientific process of creation.[38] To teach only evolution in schools, Bird argued, impinged on the rights of creationist students: "Exclusive public school instruction in the general theory of evolution abridges free exercise of religion."[39]

Bird and the other "scientific" creationists of the Institute for Creation Research (ICR) in San Diego promoted this strategy as a way to insist on the teaching of two theories in public-school classrooms. Along with ICR founder Henry Morris, he crafted a model equal-time resolution for school districts. The ICR also sponsored projects such as a model two-theory science curriculum. The curriculum's author, Richard Bliss, had served as both the director of curriculum development for the ICR and as science consultant to the Racine, Wisconsin, school district. Students using Bliss's curricular materials read that "the creation model, in its modern sense, has been developed scientifically by scientists who interpret present scientific data about life as the result of *original design*."[40]

Bliss's repeated use of the word *science* was no mistake. Like Bliss, ICR activists insisted that their campaign was based on science, not religion. Nevertheless, just as it had in the 1920s, opposition to evolution in the last decades of the twentieth century usually meant more than simply a scientific disagreement with the theory of evolution. As judicial decisions in the 1960s and 1980s recognized, creationists hoped for both protection and propagation of their religious beliefs. Though ICR leaders insisted that creation science could be taught for its scientific merit alone, they also didn't make much of a secret of their hopes for the evangelical power of scientific creationism. Morris, for instance, insisted that public schools represented "the most strategically important mission field in the world."[41]

Although Bird, Bliss, and Morris preferred to persuade school boards one by one, other evolution opponents took a more aggressive approach. Catholic creationist Paul Ellwanger, the public face of the creationist Citizens for Fairness in Education, pushed a version of Bird's resolution as a model state bill. Two states quickly passed such laws, and over twenty states considered them.[42]

The first of these new laws, Arkansas' Act 590 of 1981, echoed Bird's call for "Balanced Treatment of Creation-Science and Evolution-Science." The law emphasized that such treatment would "protect academic freedom . . . [and] freedom of religious exercise."[43] Despite this new emphasis, the legislation did not convince US District Court judge William Overton. After a two-week trial that brought a predictable parade of scientists, scholars, reporters, and gawkers to Little Rock, Overton ruled that the law would push the state to "an excessive and prohibited entanglement with religion." As he found the law lacking a primarily secular purpose, he ruled it unconstitutional.[44]

The second equal-time law, Louisiana's 1981 Balanced Treatment Act, had a similarly spectacular but ultimately unsuccessful career. By the time the law reached the US Supreme Court in the case of *Edwards v. Aguillard* in 1987, Wendell Bird, the ICR, and other evolution opponents had clashed with the American Civil Liberties Union, the National Association of Biology Teachers, the National Science Teachers Association, and the American Association for the Advancement of Science over the propriety of the two-model approach. The Supreme Court agreed that the Louisiana law had a primarily religious purpose and was therefore unconstitutional.[45]

As evolution supporters had done in the aftermath of the Scopes trial in 1925, some mainstream scientists hailed this decision as the final triumph over evolution opponents.[46] However, just as the Scopes trial had failed to provide the lasting solution that many had hoped for, this round of struggle over evolution education had mixed results. Although evolution supporters had convinced judges of the impropriety of equal-time laws, they hadn't convinced the general public. The case for including both evolution and creation in public-school science classes remained enormously popular. Politicians and their constituents supported a two-model approach in significant majorities. On the campaign trail in 1980, presidential candidate Ronald Reagan publicly endorsed a two-model vision. If evolution were to be taught in America's schools, he asserted, "then I think the biblical story of creation should also be taught."[47] In 1981, an Associated Press-NBC News poll found that 76 percent of respondents thought that public schools should teach both "theories."[48] In 1999, a Gallup poll found that 58 percent of American adults were in favor of the two-model approach.[49]

In an attempt to overcome judicial opposition to religiously inspired anti-evolutionism, evolution opponents soon offered a new idea, "intelligent design," which we'll consider in chapter 6. The idea that the complexity of life could not be explained by "blind" or undirected natural processes was a very old one, but the modern incarnation dated to a 1984 publication by Charles B. Thaxton, Walter L. Bradley, and Roger L. Olsen, with a foreword by Dean H. Kenyon.[50] In 1989 Kenyon, a professor of biology at San Francisco State University, collaborated with Percival Davis, a community-college biology teacher and successful textbook author, to produce a new curriculum supplement, *Of Pandas and People*, which explicitly used the term *intelligent design* to explain the origins of life.[51]

Kenyon and Davis had watched legal developments closely. Although they admitted that their motivations for their new textbook were largely religious, they decided to revise its content to emphasize secular scientific arguments.[52] Their point was simple: natural processes are not the most plausible explanation for the incredibly complex mechanisms of living beings.

Other intelligent-design advocates had a less pronounced religious motivation. Michael Behe, a biochemist at Lehigh University in Pennsylvania, embraced the notion of design as the best solution to the scientific questions that continued to challenge biologists. Behe argued that some organic forms showed such "irreducible complexity" that random mutations could never have created them. In brief, he believed that irreducibly complex structures could not have evolved; if any one part were missing, such partial structures would not offer any evolutionary benefit. The best explanation, he insisted, was that these irreducibly complex structures are the products of the hand of an intelligent designer.[53]

Behe's public advocacy of intelligent design later prompted Lehigh University's Department of Biological Sciences to post a disclaimer of his work: "While we respect Prof. Behe's right to express his views . . . it is our collective opinion that intelligent design has no basis in science, has not been tested experimentally, and should not be regarded as scientific."[54] Other institutions have proved more welcoming. In 1990 Bruce Chapman, a prominent politician in the state of Washington and founder of Seattle's Discovery Institute, dedicated one part of the institute's work to the overthrow of "scientific materialism." Since that date, the Discovery Institute has provided an institutional base for many intelligent-design advocates.[55]

Just as other evolution opponents had since the days of John Scopes, intelligent-design activists attracted the most attention when their ideas became the subject of a high-profile court case. In 2004, the school board of Dover, Pennsylvania, voted to include intelligent design explicitly in ninth-grade biology classes. An anonymous donor soon gave sixty copies of *Of Pandas and People* to the high-school library.[56] The trial in 2005 gave intelligent-design advocates and foes a chance to make their cases, once again under an intense media spotlight. In the end, judge John E. Jones III ruled that intelligent design violated the requirement that classroom instruction have a primarily secular purpose. Judge Jones also argued that using intelligent-design curricula would put the public schools in the unconstitutional position of endorsing a religious idea.[57] To add further insult, the citizens of Dover soon voted out the school board members who had supported the inclusion of intelligent design in their schools.[58]

In this case, as in other late-century disputes over evolution education, intelligent-design advocates presented themselves as the side of the shut-out minority, closed out of mainstream science by closed-minded and high-handed scientific Brahmins. In the introduction of *Of Pandas and People*, for example, Kenyon and Davis described their work as an attempt to "balance the biology curriculum." Unlike dogmatic supporters of an evolution-only curriculum, they argued, their work pushed students to weigh a variety of ideas and arguments about evolution and the origins of life.[59] Similarly, Behe published a blistering response to the ruling in *Kitzmiller v. Dover*. The court in that case, he insisted, had been overly influenced by "the presumptions and prejudices of the current group of practitioners" of science. Instead, science should be defined as the "unrestricted search for the truth about nature based on reasoning from physical evidence." By that unbiased definition, Behe reasoned, "intelligent design is indeed science."[60] Other intelligent-design advocates attacked mainstream scientists as relying on "propaganda"[61] or teaching evolution as "unquestionable dogma."[62]

In Dover, Pennsylvania, in 2005, as they had across the country since 1960, evolution opponents insisted on their rights as beleaguered minorities. In sharp contrast to the first evolution-education controversies of the twentieth century, they didn't hope to have evolution banned outright. They didn't insist that evolution be prohibited from public schools in order to protect the faith of students. On the contrary, these mostly religious activists insisted that they didn't act for religious reasons. Evolution, they claimed, represented questionable science. Only by including scientific objections to evolution could public schools teach the subject fairly and adequately.

As was the case with their counterparts elsewhere, Dover's anti-evolution arguments failed to convince the courts. To those familiar with the long history of conflict over evolution education in the United States, it seems clear that this ruling will not settle the debate one way or another. Indeed, though court decisions have tended to reject the claims of "scientific" creationists and intelligent-design advocates, the decision in the court of public opinion has been much harder to sway or to trace. Politicians and activists have continued to score successes in keeping various anti-evolution ideas afloat. From 1999 to 2005, for instance, the state board of education in Kansas swung back and forth in its support for limiting evolution education in Kansas schools.[63] In the US Senate, discussions of the controversial federal education legislation, the No Child Left Behind Act of 2002, included debates over the need to teach a range of scientific theories about evolution and creation.[64]

The recent career of evolution in actual classrooms has been even harder to discern. In a 2007 survey of public high-school biology teachers, only 28 percent claimed to teach evolution in a way consistent with the suggestions of mainstream scientists and science education professional organizations. At the other extreme, this survey found that 13 percent of respondents explicitly taught either creation "science" or intelligent design. In the wide middle, roughly 60 percent of respondents taught evolutionary theory in an ambivalent way. Some reported teaching both evolution and creationism. Others said they sought to downplay the topic of evolution entirely in order to avoid controversy.[65] In short, the best evidence from actual classrooms is that most twenty-first century students do not receive much education about the idea of evolution. Like their counterparts throughout the twentieth century, educators today largely prefer to avoid the controversy over the subject of evolution education, which is still alive and well.

What lessons can we draw from this decades-long history of controversy over the teaching of evolution? As we have seen, the fortunes of evolution supporters and opponents have basically reversed over this stretch of time: in the early days, supporters fought to get a foot in the educational door; now opponents fight to keep that door from closing behind them on their way out. This is a profound change, which perhaps explains some of the bitterness of the current dispute. In any case, we have yet to address the central philosophical issues raised by the controversy. We do that in the next four chapters. Along the way, we offer our curriculum and policy recommendations: what should public schools and their teachers do to manage this longtime controversy in the classroom? We will argue that seeing opposition to evolution instruction as an episode in the long history of religious and cultural dissent in US education, as the historical record suggests we must, opens the door to a science education with the potential to accommodate the legitimate concerns of both evolution supporters and evolution opponents.

In short, we propose a simple strategy: public schools must separate out the goal of teaching so that students achieve knowledge and understanding of evolutionary theory from the goal of fostering any sort of belief in the theory or its religious or philosophical implications. Conveying an understanding of evolutionary theory is a clear duty for public schools. Advocating any sort of religious belief or belief that challenges existing religious belief, however, is clearly not. Though this proposal seems simple, it raises some difficult philosophical issues that the coming chapters will address.

CHAPTER 5

Evolution, Creation, Science, Religion, and Public Education

Throughout the history of evolution versus creationism controversies, both sides have claimed to be not only the side of true science but also the side of better science. Indeed, both sides have ridiculed the pretensions of the opposition as nonscientific, blinded by ideological assumptions and petty peer pressure. Given the persistent debate over scientific credentials and superiority, perhaps the most fundamental question in the controversy concerns the scientific status of both evolutionary theory and "creation science" and intelligent design (ID). What entitles a theory to the designation "scientific"? Can scientific theories be distinguished from their nonscientific brethren? If so, how? While advocates of evolution and of creationism or ID have defended their theories as scientific, both theories have also been decried by their critics as religious. So a related set of controversies concerns the propriety of the distinction between science and religion. Is creation science really just thinly disguised religious doctrine, as its critics sometimes claim? Similarly, is evolutionary theory "just another religion," as *its* critics allege? The answers to these questions will help us determine whether they are entitled to equal treatment in the science classroom. In this chapter, we discuss evolution and creationism; we consider ID in the following chapter.[1]

WHAT IS A SCIENTIFIC THEORY?

Philosophers of science used to agree (insofar as philosophers agree on anything) that scientific theories could be cleanly distinguished from their nonscientific counterparts; they agreed as well about the criteria by which these theories could be so distinguished. The situation has changed in recent decades. In this chapter, we first lay out the earlier, "standard" view and apply it to the evolution versus creationism controversy. We then consider the current view that there doesn't appear to be a sharp distinction between science and nonscience, and assess the impact of this shift on our assessment of the controversy.

The standard view holds that several criteria together determine the status of a theory or hypothesis as scientific or not scientific. To be scientific, a theory/hypothesis must be open to empirical confirmation or disconfirmation; it must yield testable predictions; it must explain; it must be held nondogmatically; and it must be tentative, fallible, and self-correcting. Most important, a theory/hypothesis is scientific if and only if it is *testable*: *confirmable/verifiable*, or *disconfirmable/falsifiable*.[2] That is, a genuine scientific hypothesis must yield predictions that can in principle be determined to be correct or incorrect, and can count in favor of or against the hypothesis under consideration. To take an overly simple example: "Copper conducts electricity" counts as scientific, because we can pass current through pieces of copper and examine its conductive powers, that is, how completely and efficiently the current passes through the copper. On the other hand, "Copper is composed of atoms and souls" does not count as scientific, because although we can study the atoms, we can neither confirm nor disconfirm the existence of the postulated souls. On the standard view, then, all legitimate scientific hypotheses are verifiable/falsifiable, at least in principle.[3]

In addition, scientific theories/hypotheses have *predictive* and *explanatory* powers: the testable predictions they produce can explain relevant phenomena. First, scientific theories must have *predictive power*. In particular, scientific theories/hypotheses should be able to predict[4] novel observations. One of the great triumphs in the history of science was the prediction, made independently by John Couch Adams and Urbain Le Verrier, of the existence and location of Neptune on the basis of irregularities in the orbit of Uranus. Einstein's prediction of the effect of gravity on light rays, briefly discussed below, is another example. Scientific theories must also have *explanatory power*: they must explain why the facts within the purview of the theory are as they are and are not some other way. A theory of celestial mechanics must be able to explain why, for example, Venus and Mars follow the paths they do and not any other paths.

According to the standard view, scientific theories/hypotheses are also *tentative* and *self-correcting*. That is, they are never taken to be "the last word"; they are always open to disconfirmation in light of future evidence and hence tentative. Moreover, since negative evidence forces us to recognize difficulties with tested hypotheses, we can improve them. In this way, science is *self-correcting*: it improves, at least ideally, by learning from its failures and formulating theories/hypotheses that better handle the evidence. (It is worth noting that strictly speaking, these are features of *scientists* rather than of theories/hypotheses. We will not tarry on this point here.)

Although this discussion of criteria for evaluating scientific theories is hardly complete, it will suffice for present purposes. Although there are other important criteria, on the standard view, as we have seen, a theory is a legitimate or genuine scientific theory only if it meets the criteria of testability, explanatory power, and predictive power.

Consider a famous example: "Light rays bend when they pass through the gravitational field of massive objects." This hypothesis is eminently testable. We need only look and see (with appropriate apparatus) whether light waves *do* in fact "bend" when they pass near such objects. This hypothesis is a testable implication of Einstein's theory of general relativity, which predicts the bending of light rays as they follow the curvature of space in the vicinity of massive objects. When Einstein formulated this prediction, it was not possible (technologically) to test it. Scientists had to wait some years for a solar eclipse in order to conduct the test. When the test was carried out in 1919, it confirmed Einstein's theory. Light emitted from a conveniently located star did indeed bend as it passed near the sun—by just about as much as Einstein had predicted it would.[5] That is, the observed evidence confirmed the hypothesis. But it was possible that the test could have shown that light did not bend as Einstein had predicted. In that case, the evidence would have disconfirmed the theory. So the theory was testable and disconfirmable/falsifiable. And, of course, it still is. New evidence could be uncovered that would disconfirm, falsify, or to some degree or other count against the theory of general relativity. So this hypothesis is clearly a scientific one, on the standard view.

How does evolutionary theory fare, according to the standard view? Is it scientific?

IS EVOLUTIONARY THEORY A SCIENTIFIC THEORY?

What is the theory of evolution? What does it say? It would take a different book to answer these questions adequately, but here we can give only a brief summary. So, briefly: according to the theory, all living organisms on earth are descended from previously existing organisms. They are related to their ancestors genetically: offspring inherit the genetic material of their parents.

This genetic material is usually replicated accurately, but occasionally errors occur, such that the genes of the offspring are different, "incorrect" or imperfect copies of the parental genes. These alterations, or changes, are called *mutations*. Many mutations are inert and don't materially affect the organism's chances of survival and reproduction. Some mutations are harmful, in the sense that they render the offspring less likely to be successful at survival and reproduction and so the passing on of their genes. Still other mutations are helpful in that they increase the organism's fitness, enhancing its ability to survive, reproduce, and pass on copies of its genetic material to its offspring. In the environment into which the organism is born—its *ecological niche*—it must, if it is to survive and reproduce, avoid predators, diseases, and other threats and compete with its fellow species members for food and mates. In this "struggle for survival," organisms with particular traits will be more successful than their peers at passing on their genetic material. The process of *natural selection*—in which fitness-enhancing traits determined by that genetic material are "selected for" by virtue of more successful survival, reproduction, and gene replication and transmission compared to fellow organisms without those traits—can be thought of in terms of *blind variation and selective retention*:[6] the mutations are unguided, undirected errors in the copying of genes, most of which are inert or neutral, some of which are harmful, but some of which are fitness-enhancing; the latter are likely to be selectively retained in that they spread through the relevant population of organisms over the course of succeeding generations precisely because they increase the likelihood of the organism's survival and reproduction.

Several points are worth emphasizing. First, the evolutionary process is undirected: mutations occur randomly, unguided by any "evolution director" and not aimed at any end point. There is no ideal end point at which the evolutionary process aims; fitness is relative to niches, and an organism or population that is well adapted to its present niche may find itself poorly adapted when its niche changes (for example, when the environment becomes hotter or colder, wetter or drier, lighter or darker, or when a key food source is lost or a new predator introduced). Second, whether a given mutation is inert, harmful, or helpful is a matter of the niche in which the bearer of the trait resulting from that mutation finds itself: a mutation that confers improved vision, for example, will be selected for only if better vision in fact enhances fitness. (So for creatures that live and die in dark caves, for example, enhanced vision will not enhance the chances of survival and reproduction, and may harm those chances if the enhancement of vision comes at the cost of reduction of other, more important-for-survival traits, such as hearing and smell, that might improve the ability to detect food, predators, or potential mates.) Third, this far too brief account of the theory leaves out many details, and ignores several problems and complicating factors. For example, natural selection is perhaps the major but not the only evolutionary phenomenon leading to the change of frequencies of various mutations; genetic drift is another. An open question is whether natural selection operates at the level of the gene, the organism, or the population: which is selected for? This is the "unit of selection" controversy. These and other issues are well worth pursuing but cannot be taken up here; some will be addressed in subsequent chapters. We hope that this brief account of the theory will suffice for consideration of its scientific status.[7]

Is the theory of evolution, as just briefly characterized, scientific? As we have seen in previous chapters, evolution opponents often claim that the theory is in fact not scientific, that it fails to meet relevant criteria for being scientific, and that it is actually nothing more than a dogma, assumption, or presupposition, held by its advocates as an unconfirmable article of faith. Is this correct?

To answer this question, it is helpful to distinguish three distinct senses of *evolution*. First, there are the *facts*—the empirical findings—of biology that the theory of evolution purports to explain, including the close morphological and biochemical relationships among organisms and species, the geographical distribution of related species, patterns of embryological development across species, the existence of vestigial genes and organs, the existence of anatomical and biochemical traits that are functionally related to particular environmental factors, patterns of molecular similarities among living species, genetic changes in both laboratory populations and bacterial and viral populations during the course of infection and disease, genetic responses of populations of microorganisms to chemical toxins, and genetic changes during the histories of domesticated animals and plants and in natural populations in response to ecological fluctuations.[8] Second, there is the *theory* of evolution, which purports to explain these facts, mainly in terms of natural selection but also in terms of reproductive isolation, genetic drift, and so on. Third, there are nonbiological processes of development and change, for example the "evolution" of a star, galaxy, or social/cultural group. Once these three senses of *evolution* are distinguished, the charge that evolution is a dogma or indefensible assumption is easily parried. The facts are clear, readily observable, and confirmed in laboratories and classrooms every day. The theory is indeed theoretical, and in that sense not observable; but it is very well confirmed and has enormous explanatory power, and so is not at all a dogma or article of faith. The nonbiological processes are, strictly speaking, irrelevant to this issue, because they are not biological; but our understanding of them is likewise well supported both factually and theoretically, and so they hardly qualify as dogmatic assumptions concerning the evolution of living species.

Not only is evolution not a dogma or unsupported assumption, it comfortably meets the criteria put forward by the standard view. First, evolution is testable and falsifiable. If, for example, we found no evidence of mutation, molecular/chemical relationships among the genes and chromosomes of different species, shared genetic material across species, species flourishing in favorable environments and dying out in unfavorable ones, or the several other facts mentioned in the preceding paragraph, we would have good reason to question both the truth of evolutionary theory and the very existence of the biological facts it explains. Evidence is clearly capable of falsifying or disconfirming the theory.

Evolutionary theory also has impressive explanatory power. It explains, for example, the survival and transformation of some species and the extinction of others. It explains the close biochemical relationships among species. It explains genetic continuity. In collaboration with geography, geology, and other relevant sciences, it explains the biogeographical distribution of species across the planet. It explains the complex patterns of observable similarities and differences among the many characteristics of organisms and species across the biological world. It explains all the facts mentioned in the previous paragraphs. Evolutionary theory offers a powerful explanation of the "unity in diversity" of the biological world seen by students and researchers every day.

Finally, evolutionary theory has predictive power. The interrelatedness of all life on earth is a prediction generated by evolutionary theory that is extremely well supported by the evidence. More specific predictions—for example concerning the trajectory of drug resistance in HIV/AIDS research, in which the virus reacts predictably by mutating in ways that make it resistant to the drugs in its environment (that is, the interior of the body)—have successfully guided efforts to devise more effective drug combinations ("cocktails") to treat the disease. Some of Darwin's predictions, for instance that Precambrian fossils would be found, were correct. Other predictions based on the theory, such as that we should find a "transitional" whale fossil with both teeth and baleen, were correct. Many other examples could be given. The point is not so much that many predictions based on the theory have been correct (though they have). The main point is that the theory licenses predictions, and in that way satisfies the criterion of predictability or predictive power.

In short, then, according to the criteria put forward by the standard view, evolutionary theory is indeed scientific.

Some evolution opponents have disputed the theory's scientific status. In particular, they have argued that the theory is not a legitimate scientific theory because it is not falsifiable; that no conceivable observational result could disconfirm the theory. We have already seen that this is not the case: failure to observe mutations, extinctions, molecular/genetic continuities, and the like would call into question the existence of the phenomena the theory is meant to explain, thereby tending to disconfirm the theory, as would a broad range of more specific predictive failures of the sort just mentioned, were they to occur. We could observe these things; thus, the theory is falsifiable. But we don't observe them. Our observations give us no reason to think that evolution theory is false and lots of reason to think it true.[9]

IS "CREATION *SCIENCE*" RIGHTLY SO CALLED?

We have been arguing that evolution is indeed science, at least according to the standard view. Can we say the same for creationism? No. "Creation science" is a misnomer. Whatever else creationism may be, according to the standard view it is *not* science. That is, it fails to meet the criteria we have been considering: it is not testable (or, as we will see in a moment, is testable and has been falsified); makes no serious predictions (or makes them but they are false); and offers no explanations. It also has no research agenda and makes no contribution to the biological literature. Given all this, how could it possibly count as science?

Most tellingly, creationism fails to meet the standard view's criterion of testability/falsifiability for the simple reason that *no possible test result could be incompatible with it*. A creationist can counter any conceivable observation or finding by saying, "The Creator created things that way." For example, if it turned out that primates such as humans and chimps did not share most of their genetic material, we would count that as telling against evolutionary theory, at least as currently understood. In fact, however, humans and chimps do share most of their genes. This counts in favor of their common recent ancestry as predicted by evolutionary theory and should therefore also count against creationism. But the creationist can easily parry this apparently negative result by declaring that the Creator created these species in this way. There is nothing special about this example; for any observation or research finding that seemingly tends to disconfirm creationism, the creationist can simply declare the finding to be reflective of the Creator's creation. That is, no conceivable research finding can disconfirm or falsify creationism.

It may be replied that in fact creationism is not only falsifiable but false, because the truth of evolution entails the falsity of typical creationist claims concerning the origins of species: evolution says they've evolved from common ancestors, while creationists say that at least some of them, and in particular humans, did not evolve but were rather created *ex nihilo* by God / the Creator. We have just enumerated some of the overwhelming lines of evidence suggesting that all species evolved from preexisting species, and so the falsity of such creationism. Thus, we should say that creationism is both falsifiable and in fact false.

This is a powerful point, and creates a dilemma for the creationist: either he can say about everything, including the evidence favoring evolution, that "things are the way they are because the Creator created them that way," in which case creationism is not falsifiable and so is not scientific; or he can say that "the Creator didn't create things such that evolution is true, because evolution is false," in which case creationism is refuted by the evidence for evolution. So creationism is either not science or not true. If it is granted that the creationist can't say, "Things are the way they are because the Creator created them that way" about evolution itself, because he takes that theory to be false, then he stands refuted by the powerful and wide-ranging evidence for evolution. On the other hand, if he says this of evolution, too, then his theory is unfalsifiable. Neither of these options offers much comfort to the creationist. Since we think it clear that he will not accept the horn of the dilemma according to which his theory is false, we will proceed on the assumption that he embraces the other horn, according to which it is unfalsifiable.[10]

Similarly and relatedly, creationism lacks predictive power. Because no conceivable test result could falsify it, creationism can make no genuinely testable prediction. Consider: it would be nice for creationism if it could successfully predict a lack of shared genes across species, since that would make it look like the Creator created each species *de novo*. But creationism cannot make this prediction, given that, as we have seen, any possible distribution of genes across species would be consistent with creationism, because the Creator could have created species that way. Again, the point is perfectly generalizable. Because any predictive result can be reconciled with creationism by declaring that result to be the way the Creator created things, creationism has no predictive power. No prediction can turn out to be incorrect, because creationism licenses them all as the way the Creator created. On the other hand, as we have seen, if it does not so license all predictions, such that some of them—for example, that concerning the proportion of shared genetic material across species—can be put to the test, then those tests have already been carried out and the predictions have failed. So either the theory lacks predictive power and so fails to meet the criterion of predictive power, or it is simply false.

Similar points can be made with respect to explanation and explanatory power. To have such power, a theory must be able to say why things are the way they are and *not some other way*. Creationism cannot do this, because, as we have seen, any way things turn out to be can be declared to be the way the Creator created them. Because "the Creator created things that way" can be offered as an explanation of every conceivable state of affairs, creationism in fact explains none of them. Creationism is compatible with all possible findings, and so explains none: to explain a given finding or phenomenon, an explanation must be incompatible with at least some possible findings. (I don't explain today's overgrown state of my backyard in terms of the "fast-grow" chemical I applied last month if I could also explain tomorrow's barren and bedraggled state in terms of that same chemical. It does not explain the one if it also "explains" the other.) Since "the Creator created things that way" is the creationist's explanation for every possible observation or experimental finding and so cannot explain why things are not the way they are not, creationism is entirely lacking in explanatory power.[11]

For these reasons, creationism—if it is not simply to be declared false—fails to count as science, according to the criteria set out by the standard view. It is not testable/falsifiable, has no predictive power, and is explanatorily inert. (The applicability of these reasons to the more recently advocated alternative to evolution, "intelligent design," is as yet unclear. We consider ID in the next chapter.)

BEYOND THE STANDARD VIEW: THE DEMARCATION PROBLEM

As we have seen, according to the standard view, science can be sharply distinguished from nonscience; and on the basis of its failure to meet the criteria of genuine science, creationism can and should be legitimately excluded from the science curriculum. That is, "creation science" is a misnomer: it is not falsifiable, has no predictive or explanatory power, and so fails to constitute genuine science; or it is simply false. However, most contemporary philosophers of science have given up on what Popper called the "demarcation problem"; they think that there are no necessary and/or sufficient conditions that suffice to distinguish the genuine article from counterfeit imposters.[12]

If there is no way of sharply distinguishing science from nonscience, does it follow that creationism in particular, or theories with (on the standard view) questionable scientific credentials more generally, are worthy of inclusion in the science classroom? The short answer is no. Even if we cannot sharply distinguish science from nonscience, we *can* distinguish between better and worse entitlements to the honorific title "scientific." Moreover, given a range of alternative theories, we can evaluate them in terms of their quality, and in doing so distinguish excellent from good, good from bad, bad from dreadful, and live from dead theories. There may not be a sharp demarcation between science and nonscience, but there is a continuous spectrum on which theories fall, ranging from excellent to poor, dreadful, or worse. To rule a view out of the science curriculum, it is not necessary to show it to be nonscience or pseudoscience. It is enough to show it to be of poor quality, clearly inferior to other theories in the same domain. And this we can do by appeal to the very criteria advanced by the standard view: testability, falsifiability, predictive power, explanatory power, responsiveness to apparently disconfirming evidence, and so on. The last of these introduces a historical element to theory evaluation: theories that at one point might readily have met criteria of scientific quality might meet them less well or come to fail to meet them altogether over time.[13] Accordingly, the important question is not "Is creationism science?" but rather "Assuming that we cannot sharply distinguish science from nonscience, which—creationism or evolution—is *better* science?" And here the answer is clear: in terms of the criteria just mentioned, evolution is clearly superior, and creationism sufficiently inferior to be justifiably excluded from the biology classroom if presented as a scientifically legitimate alternative to evolution.[14]

ARE EITHER EVOLUTIONARY THEORY OR CREATIONISM THINLY VEILED RELIGION?

When defining contested concepts like that of religion, some ineliminable controversy is perhaps to be expected. Still, in considering the place of evolutionary theory and creationism in the science classroom, some account of religion is required, given that both sides in this controversy accuse the other of being religion rather than science and so undeserving of a privileged (if any) place in the science curriculum. In full awareness of the difficulty of providing a general characterization of religion that will be acceptable to all parties in this dispute, we will take a religion to be a system of belief that has as its most prominent feature some conception of a divine creator or of a supernatural or transcendent order in the universe, and that requires the acceptance of some doctrine on faith. Such faith-based doctrine is held by its proponents to be impervious to falsification and unchallengeable by empirical evidence.

Is the state, by teaching evolution, supporting a particular religion, often called secular humanism? It is not. Secular humanism, an ethical view that affirms the fundamental dignity and worth of humankind, is essentially nonreligious. It holds no doctrine on faith; it makes no claims concerning the divine or supernatural. Secular humanists *qua* secular humanists do not pray, meditate, or acknowledge divine, supernatural, or transcendent forces in the universe.[15] Assuming, as creationists assert, that secular humanists are likely to embrace the theory of evolution, it is gratuitous to label either secular humanism or evolution a religion. Secular humanism does not share the features definitive of religious systems of thought. More to the point, neither does evolutionary theory, which is after all a theory concerning essentially biological phenomena. No part of the theory addresses matters divine, supernatural, or transcendent; none is beyond critical scrutiny or evidential challenge.[16]

Creationism, on the other hand, seems clearly enough to count as a religion according to our characterization, or at least to involve religious beliefs or precepts. It affirms a divine Creator and attributes a supernatural order to the universe. Its advocates in the United States most commonly accept Christian doctrine and moreover hold those doctrines as a matter of faith and so as beyond evidential challenge. Insofar as religion has no place in the science classroom, then, it seems clear that while creationism should on that basis be excluded from the science curriculum, evolutionary theory has a perfectly legitimate place in it.

MUST THE STATE BE "NEUTRAL"?

As we saw in chapter 4, it has often been said that the public schools should be "neutral" in this controversy: afford both evolution and creationism equal time in the science classroom, and let students decide for themselves which view to embrace. There is clearly something plausible, albeit misleadingly so, about this. We must be careful here, not least because the word *neutrality* is ambiguous. In one sense, *neutrality* means not unfairly favoring one alternative over another; in another sense, it means refraining from judgment altogether. Many advocates of creationism, including several recent highly visible US presidents and presidential candidates, suggest that the state should operate under the latter definition of neutrality. They argue that the state should present both "sides"—evolution and creationism—and let the student judge. This gravely misconstrues the role of the state in public education.

The public education system is not simply a broker of information. It doesn't simply offer students information they couldn't otherwise obtain and leave them to do with it whatever they choose. Rather, the educational system has the primary aim of *educating* students. It is not sufficient to present the alphabet or the number line to children and say, "Do with these what you will," or to offer alternative number lines, mathematical techniques, or algebraic systems and say, "Learn whichever you wish." Educators must make judgments concerning the proper and improper uses of these objects of learning. They cannot be nonjudgmental, though they can and should be unbiased, and in this sense "neutral." They must use judgment in determining virtually every aspect of the curriculum. Although they must recognize that their judgments are fallible, they must not shirk the responsibility of judging.[17]

One important criterion of this judgment is *disciplinary adequacy*. For any given discipline, educators must choose to put before students not just *any* material from that discipline but the *best available* material, as judged by those most qualified and competent to judge. Educators must uphold standards of disciplinary adequacy if their efforts to impart a sense of those standards to students are to be either justified or successful. In invoking its right to make judgments concerning curricular materials, the state is simply meeting its obligation to provide high-quality education to its citizens. Nonjudgmental neutrality is indefensible as a curricular rule.

The state, then, in favoring evolutionary theory over creationism in the science curriculum, is not failing to be neutral in the sense of the term that is relevant to this issue. It isn't being biased or unfairly favoring the former over the latter, for there are exceedingly strong reasons for favoring evolutionary theory: it meets the disciplinary standards of science; creationism does not. Moreover, the demand that the state be "philosophically neutral" is unwarranted, if "neutrality" is construed as nonjudgmental. Refraining from judgment concerning the scientific merits of both evolution and creationism and their worthiness for inclusion in the science curriculum is failing to maintain standards of disciplinary adequacy. "Anything goes" is an indefensible principle of curriculum content.

Up to this point, we have limited ourselves to consideration of the place of creation "science" in the science curriculum. This limitation must now be rectified, for creationism has largely been replaced in the alternative-to-evolution literature by intelligent design. In the next chapter, we evaluate ID's claim to a proper place in the science curriculum.

CHAPTER 6

Beyond "Creation Science": The Scientific Status of Intelligent Design

In keeping with the historical account offered in earlier chapters, chapter 5 focused mainly on creationism. But as we saw in chapter 4, the contemporary preference of those seeking to overcome or mitigate the teaching of evolution is not only creationism but more often intelligent design. According to its advocates, ID better highlights the scientific deficiencies of evolutionary theory, offers a legitimate scientific alternative to evolution, and makes no religious commitments; thus, it cannot, or at least should not, be ruled out of the science classroom for that reason. In this chapter, we evaluate ID's worthiness for inclusion in the science curriculum.[1]

WHAT IS ID?

The basic idea of ID is clear: there are complex biological phenomena that cannot be explained by evolutionary theory or one of its main mechanisms of change, natural selection ("descent with modification," or "blind variation and selective retention"). These phenomena can be explained only by the existence of an intelligent designer: a presumably conscious entity of some sort, capable of designing and bringing complex biological systems into existence. We will soon examine the properties ID attributes to the intelligent designer, and ask what properties such an entity would need to have in order to accomplish the design of such complex structures as eyes, bacterial flagella, and so on. But first let us characterize ID with a bit more specificity.

As philosopher of science Philip Kitcher characterizes it, ID consists of both a negative and a positive thesis. The negative thesis is that there are biological phenomena that aren't explainable in evolutionary terms. The positive thesis is that these phenomena can be explained and understood only, as Kitcher puts it, "as effects of an alternative causal agency, one that is properly characterized as 'intelligent.'"[2] That is, ID's designer must be a causal agent, capable of making things happen in the world. Moreover, the designer must be "intelligent." We will turn to the question of the character of such intelligence soon.

The idea that known phenomena, too complex to be thought to have arisen by natural or "unintelligent" processes, must have been brought about by an intelligence of some sort might remind the reader of William Paley's famous articulation of the "design argument" for God's existence.[3] In a widely reprinted selection from this book, usually titled either "The Watch and the Watchmaker" or "The Argument from Design" in introductory philosophy textbooks, Paley asks us to imagine ourselves walking through a heath, experiencing nothing but the sights, sounds, smells, tastes, and feels of unsullied nature, when we stumble upon a (now old-fashioned mechanical) watch on the ground. We examine it, and exploring its complex inner workings, its precise organization and the intricate interconnections of its coils, springs, and gears, structured so as to tell time, we infer that it cannot have been created by blind natural processes, but rather must have been designed and made by a watchmaker:

> This mechanism being observed (it requires indeed an examination of the instrument, and perhaps some previous knowledge of the subject, to perceive and understand it; but being once, as we have said, observed and understood), the inference, we think, is inevitable; that the watch must have had a maker; that there must have existed, at some time and at some place or other, an artificer or artificers who formed it for the purpose which we find it actually to answer; who comprehended its construction, and designed its use.[4]

As students of Philosophy 101 often realize, Paley's design argument has several problems (several of which were already anticipated by David Hume in his *Dialogues concerning Natural Religion*[5]), the most important of which is that the inference to a designer (watchmaker) is licensed mainly by the lack of an alternative explanation: we think the watch must have been designed because we cannot imagine any other plausible way it could have materialized, can point to no plausible alternative process by which it could have come into existence. Were we to have such an alternative, the inference to a designer would be weakened. In the case of the watch, the inference remains secure; we have no idea how watches might come into existence other than by the activities of watchmakers. But of course Paley wasn't arguing just that watches result from the design and implementation of the design plans of watchmakers; he was arguing, by analogy, that just as the watch must have been designed, so must complex biological systems—the eye, the brain, the central nervous system, and indeed the world itself and all its parts ("all the organized parts of the works of nature"[6])—have been designed by an intelligent designer. And while we cannot readily imagine an alternative to a watchmaker, we can imagine an alternative to a designer of complex biological systems: that is, evolution, driven mainly by natural selection.

In what ways, if any, is ID superior to Paley's original case for an intelligent designer? The question is complicated by the fact that Paley was explicitly arguing for the existence of (the Christian) God: complex biological systems in particular, and the universe more generally, could not have arisen "by chance," but must have been designed by God:

> Upon the whole; after all the struggles of a reluctant philosophy the necessary resort is to a Deity. The marks of *design* are too strong to be got over. Design must have had a designer. That designer must have been a person. That person is God.[7]

The advocate of ID does not want to incorporate this claim into the version of the view suitable for inclusion in the science curriculum, though, for ID is supposed to be strictly scientific with no religious or supernatural components or implications. In which case: what must the intelligent designer be like, if he/she/it is not presumed to be God?

WHAT MUST THE INTELLIGENT DESIGNER BE LIKE? WHAT CAUSAL PROCESSES DOES THE DESIGNER INVOKE?

Here is a crucial issue for the advocate of ID. For ID to be a serious scientific alternative to evolution, the theory must provide an account of the designer and (traditionally His) intelligence, such that it provides a testable account of the causal mechanisms by which the designer brought about the biological phenomena in question. There are three options open to advocates of ID here: articulate such a testable account; articulate an account that is not testable, because it relies on religious, spiritual, or related characteristics of the designer; or remain silent. As a matter of fact, advocates have not done the first.

They are loath to do the second, because their aim is to produce an alternative to evolution that is equally scientifically respectable and so deserving of a place in the science curriculum. So they in fact have opted for the third: the target phenomena can be explained by their having been brought about by intelligent design, they suggest, but they tell us nothing about the causal mechanisms in play or the nature of the designer that implemented them.[8]

As long as the character of the designer and his/her/its intelligence is left unarticulated, ID cannot be regarded as a serious scientific alternative to evolution. Imagine the following dialogue in the science classroom:

STUDENT: How exactly does ID account for the subtle marvel that is the human eye?

TEACHER: ID holds that the eye could not have come about by way of evolutionary processes, in particular the process of natural selection, but must have been designed by an intelligent designer.

STUDENT: But how exactly? Evolutionary theory proposes a small handful of causal processes, most important that of natural selection. It is of course true that in the case of the eye, the process would have been exceedingly complex and taken a very long time, and that many steps of the process remain unknown. But at least in principle the process is understood, and future research may well enhance our present understanding of its many steps. How does ID propose that the process came about?

TEACHER: Well, the designer designed it.

STUDENT: But how? What causal process(es) did the designer make use of? Is there any evidence that these processes were actually invoked? Is there a way to test this account of the coming into existence of the eye?

TEACHER: ID does not identify specific causal processes. . . .

The point should be apparent: the teacher cannot answer the student's questions without appealing to characteristics of the designer and the causal processes the designer invoked in bringing his/her/its design to fruition—characteristics and processes that at least ideally are amenable to scientific examination. But ID specifies no such characteristics or processes. It is content to point to difficulties with the evolutionary account, in particular its alleged improbability. (We will consider this alleged difficulty below.) Absent an articulation of such characteristics and processes, it is difficult to see how ID can be considered a legitimate scientific alternative to evolution, deserving of inclusion in the science curriculum. It seems rather to offer the shell of a causal story, to be filled in, if at all, by religious/supernatural, likely Christian, doctrine. But any such filling in is done offstage and is not part of the official ID position.

IS ID SCIENTIFIC? IF SO, IS IT A BETTER SCIENTIFIC THEORY THAN EVOLUTION?

As noted in chapter 5, we should perhaps avoid trying to answer the first of these questions in terms of a sharp science/nonscience distinction; we ought not to presuppose a solution to the demarcation problem. Let us grant, then, at least for the sake of this discussion, that there is no uncontroversial basis for declaring ID to be nonscience. We then must face squarely the second question: how does ID compare, scientifically speaking, to evolution?

There are several points to make here. First, we should grant to the advocates of ID that there are many biological phenomena that we don't yet fully understand and that evolutionary theory has yet to explain. Evolutionary theory is very much a work in progress. It has made enormous progress, but there is much more for it to do. There certainly are phenomena that it cannot yet fully explain. There is nothing special about evolutionary theory in this regard; every domain of science is a site of ongoing inquiry in which scientists strive to address unanswered questions and provide explanations of currently unexplained phenomena.

Second, and despite the previous remark, at least some of the phenomena that advocates of ID point to as things evolution cannot explain in fact have been or at least in principle can be explained in evolutionary terms.[9] The claim that evolution cannot, even in principle, explain these things is simply false.

Third, granting the advocate of ID that there remain outstanding problems for evolutionary theory, it remains nonetheless (as we saw in the previous chapter) that evolution is testable and enjoys considerable explanatory and predictive power. Evolutionary theory is *better* science than ID. We say this because:

Fourth, whatever the scientific quality and status of evolution might be, ID has little to recommend it as high-quality science. As we have already seen, its characterization of intelligence is effectively empty; that is, it provides no causal mechanism that explains any of the biological phenomena that remain beyond the current explanatory power of evolutionary theory. And like creationism, ID is not testable because, absent a serious account of the characteristics of and the causal mechanisms and processes employed by the intelligent designer, it is compatible with all conceivable test results and so can be disconfirmed by none of them. Moreover, absent such an account, it has—just as with creationism, and for the same reasons as those canvassed earlier—no predictive or explanatory power of its own.[10] Consequently, to the extent that ID constitutes genuine science, it is at best "dead" science:

> If Intelligence is (unofficially) taken as a surrogate for God, we are unable to understand its aims and capacities. No principles about its direction or about its powers can be stated.
>
> Without such statement, however, intelligent design has no positive doctrine. Without positive doctrine, it provides no explanations. Without explanations it has no standing as a scientific alternative to Darwinism. And so the status claim made on its behalf is unwarranted. It is empty. Dead.[11]

IS ID NONRELIGIOUS?

Here we can be brief. Insofar as ID makes no appeal to spiritual or supernatural characteristics or processes, there is no reason to regard it as religious, either overtly or covertly. But if it makes no such appeal, there is little to recommend it as a serious scientific alternative to evolution. On the other hand, it is difficult to see how ID might enhance its testability or explanatory and predictive power except by invoking such characteristics and processes, which would not only render it religious but ironically fail to enhance its scientific credentials.

THE IMPROBABILITY OF EVOLUTION

Finally, let us consider what is often taken to be ID's greatest strength (and evolution's greatest weakness): the alleged extreme improbability of the sequence of events posited by the Darwinian account. The point is often made in terms of "irreducible complexity":[12] particular natural phenomena are so complex that the probability of their having arisen by chance is vanishingly small. The idea is clear: if the evolutionary story is so incredibly improbable, the explanation of such phenomena in terms of intelligent design seems the only credible alternative.

This improbability claim is familiar from the literature advocating ID.[13] It also plays a key role in philosopher Thomas Nagel's "Public Education and Intelligent Design."[14] Nagel's paper is addressed mainly to the legal question concerning the legitimacy of teaching ID in the public schools; but since this question (as we have seen) is intimately bound up with the scientific status of both ID and evolutionary theory, it is not surprising that it plays a key role in his discussion.[15]

Nagel claims that critics of ID tend "to overstate the legitimate scientific claims of evolutionary theory,"[16] and that "the campaign of the scientific establishment to rule out intelligent design as beyond discussion because it is not science results in the avoidance of significant questions about the relation between evolutionary theory and religious belief, questions that must be faced in order to understand the theory and evaluate the scientific evidence for it."[17] Since we have already given up the "ID is not science" thesis in light of the difficulties faced by demarcation, we ignore here Nagel's determined attempts to show that ID is in fact as entitled as evolutionary theory to scientific status. Instead, we focus on the improbability claim and its alleged implications.

On Nagel's view, "ID . . . is best interpreted not as an argument for the existence of God, but as a claim about what it is reasonable to believe about biological evolution if one independently holds a belief in God that is consistent both with the empirical facts about nature that have been established by observation and with the acceptance of general standards of scientific evidence."[18] So understood, ID's strength as a scientific alternative to evolution will depend on one's prior religious belief. For a believer—and we should note, Nagel declares himself not to believe[19]—the scientific case for evolution can legitimately seem weak, and such a believer can with good reason believe that evolutionary theory faces deep difficulties. Indeed, it seems that way to Nagel, a self-declared nonbeliever and nonscientist:

> One of the disturbing things about the public debate is that scientists engaged in it sometimes write as if the idea of fundamental problems with the theory (as opposed to problems of detail in its application) were unthinkable, and that to entertain such doubts is like wondering whether the earth is flat. This seems to me, as an outsider, a vast underestimation of how much we do not know, and how much about the evolutionary process remains speculative and sketchy.[20]

One crucial thing we do not know, he urges, is the answer to the following question: "Are the sources of genetic variation uniformly random or not?"[21] He cites Michael Behe in support of the claim that the probability that they are indeed uniformly random is "vanishingly small":

> Alterations to DNA over the course of the history of life on earth must have included many changes that we have no statistical right to expect, ones that were beneficial beyond the wildest reach of probability. Over and over again in the past several billion years, the DNA of living creatures changed in salutary ways that defied chance.[22]

Nagel does not argue that Behe is correct. Indeed, he admits that he cannot himself evaluate this probabilistic claim.[23] Rather, his point is that this doubt about evolutionary theory is scientifically legitimate, and that as long as it is an open question whether the sources of genetic variation are indeed random, the exclusion of ID from the curriculum on this basis is a mistake: it cannot be excluded as nonscience, and the openness of the question "call[s] for the consideration of alternatives."[24] Behe, for his part, is clear about the alternative he favors and defends: "The elegant, coherent, functional systems upon which life depends are the result of deliberate intelligent design."[25]

Let us grant for the sake of argument that the question Nagel points to concerning the uniform randomness of genetic variation is indeed open. What follows? There seem to us to be four important points to make here. First, contrary to Nagel's suggestion,[26] evidence for the improbability of random processes to account for the current state of the biological world is not itself evidence in support of ID. The only way it could so count is for these two options to exhaust the range of alternatives, such that if natural selection proved inadequate, ID would be the only available alternative account of the biological facts.[27] There is no reason to think this. Second, as Nagel himself emphasizes, he is arguing not that the Darwinian account is incorrect but just that the question remains open. Given the remarkable advances evolutionary theory has made in its relatively young life, as well as biologists' more or less uniform confidence[28] in its future explanatory potential, it seems to us unduly pessimistic to regard that potential as so seriously weak as to demand the acceptance of such a flawed, explanatorily empty alternative—because third, as we have seen, the alternative on the table, ID, remains merely the shell of a scientifically respectable alternative to evolution, in that it offers no alternative testable causal accounts of the mechanisms in play in allegedly "intelligent" design.

Fourth, and most important: the alleged improbability of complex biological phenomena having arisen by chance is itself dubious, because calculations of such improbability themselves rest on suspect assumptions. In fact, the concern over improbability has been addressed and rejected many times over. Kitcher's discussion is once again instructive. He distinguishes two distinct arguments used by advocates of ID to establish the extreme improbability of processes of natural selection producing the actual phenomena: the "concrete case" argument, which "selects a collection of evolutionary changes, discusses them in detail, then endeavors to show that there's no conceivable process of natural selection that could have started from the original group of organisms and culminated in the finally modified group"; and the "computational" argument, which "abstracts from the details of individual cases, and presents them in terms of a more skeletal description, which is deployed to assign some basic probabilities. This basic assignment then allows for a calculation of the probability that the transition could have come about through the action of natural selection, and, because the resulting estimate is extraordinarily tiny, the conclusion is that causation by natural selection is, to all intents and purposes, impossible."[29] Let us consider each in turn.

The most well-known advocate of the concrete case argument is Behe, who argues that once we attend to the relevant biochemistry, the complexity of even apparently simple structures, like flagella and cilia (hairlike appendages that enable some bacteria to move), is far greater than can be explained by natural selection. In a frequently cited (including by Kitcher) summary passage, Behe writes: "As biochemists have begun to examine apparently simple structures like cilia and flagella, they have discovered staggering complexity, with dozens or even hundreds of precisely tailored parts. It is very likely that many of the parts we have not considered here are required for any cilium to function in a cell. As the number of required parts increases, the difficulty of gradually putting the system together skyrockets, and the likelihood of indirect scenarios plummets."[30]

The problem, as Behe articulates it, is indeed formidable: if "dozens or even hundreds of precisely tailored parts" are required to assemble a functional cilium, and if natural selection must produce each randomly and separately, and if they must all be present in an organism at the same time in order to be assembled, the probability of all that happening is indeed vanishingly low. But is this the right way to estimate the probability of natural selection producing a functioning cilium? One reason to think not is that the problem is posed in terms of independent, unconnected parts coming together, rather than in terms of increasingly effective functional properties. Darwin's own explanation of the evolution of the eye, in terms of increasing sensitivity and ability to respond to light, illustrates this more tractable way of construing the problem.[31] Rather than having to identify the causal processes by which a large number of independent events occurred and "came together," the challenge for the advocate of natural selection is rather that of explaining the selective forces that brought immobile or less mobile bacteria, by a series of stages, to the structures of motion (cilia and flagella) that we observe today.

Can contemporary biologists do this? No—at least not completely. There is at present insufficient comparative knowledge of the genes (and associated proteins) of bacteria with and without appendages of motion. But such knowledge is manifestly not beyond the reach of contemporary biochemistry—it exists to a considerable degree for the eye—and it is unduly skeptical to conclude, from our current inability to explain the evolution of flagella, that their existence is inexplicable in terms of natural selection and must instead be explained by appeal to an intelligent designer. The concrete case argument does not demonstrate the fantastic improbability of explanations in terms of natural selection that it trumpets. It does point to substantial gaps in contemporary biochemical knowledge; but this is acknowledged by all sides, and does not in the least support the conclusion that natural selection cannot, even in principle, explain the complex phenomena of which instances of the concrete case argument make so much.[32]

What about the computational argument? Behe also offers this one,[33] as do many other advocates of ID. The problem here is that our ignorance of the relevant initial conditions—concerning both relatively specific phenomena such as the evolution of particular biochemical mechanisms or the assembly of specific proteins, or the more general and fundamental one of the origin of life—is so great that there is no meaningful way of assigning the relevant probabilities. How, for example, can we estimate the probability of a replicating molecule arising in the "primordial soup," given our lack of knowledge of the ingredients of that soup? Advocates of the computational argument offer such probability estimates, but in the present state of biological knowledge these estimates are essentially arbitrary. As long as that is our situation, the computational argument offers us no reason to despair of the prospects of evolutionary theory.[34] And this is especially true in light of our earlier consideration of the lack of an account of the causal mechanisms employed by the intelligent designer offered by its alleged rival, ID.

This lengthy discussion hardly does justice to the vast literature and wide-ranging issues addressed in this chapter. Still, we hope we have made clear that evolution does indeed have a legitimate place in the public-school science classroom, and that creationism/ID does not. It is appropriate, of course, for science teachers to discuss difficulties with and challenges facing evolutionary theory—so long as they are genuine difficulties and challenges, as judged by standards of disciplinary adequacy noted in chapter 5—including those, just addressed, concerning improbability pointed to by advocates of ID. But it is important that insofar as ID is discussed in the science classroom—which we are not endorsing in light of its failure to meet those standards—its scientific caliber be made clear. Perhaps it is best not to label ID as pseudoscience, given the difficulties we have canvassed concerning the demarcation problem, but rather to characterize it as exceedingly poor or dead science, a "theory" that offers no substantive account of either the causal mechanisms available to the intelligent designer or the character of his/her/its intelligence.

We turn next to consideration of the appropriate aims of science education and the legitimate constraints under which it must operate. In particular, we ask: should science teachers aim at imparting *beliefs* to their students, especially when such belief, such as belief in evolution, might conflict with students' deeply held religious beliefs?

CHAPTER 7

Science Education: Aims and Constraints; Belief versus Understanding

One reason the evolution/creation debate is so contentious is that it seems to involve students' *beliefs*. Responsible parents rightly care deeply about what their children believe. If it seems to such parents that their children are forming and sustaining beliefs in school that the parents find objectionable, the latter will surely, and sensibly, object. And as our earlier chapters have demonstrated, many parents find belief in evolution—that is, belief that the theory is true[1]—deeply objectionable. So of course they object—and perhaps justifiably so. Conceived in this way, the controversy might appear to many to be unresolvable: parents who have no problem with their children learning about and coming to believe in evolution have no problem with the schools' teaching it; parents who strongly object to their children's coming to believe in evolution, most obviously because such belief is thought to be incompatible with their deeply held religious convictions, insist that the schools must not so guide their children's beliefs.

Is the assumption that both sides here embrace correct? That is, is the aim[2] of evolution education to get students to *believe* it? In this chapter, we argue against this assumption, and urge rather that the goal of evolution education should be to get students to *know* and *understand* it, leaving belief to fall where it may. While belief typically results from knowledge and understanding, it needn't; and where the teaching of evolution results in student knowledge and understanding of the theory but not belief that the theory is true, we needn't regard this as an educational failure. On the contrary, in the case in which students reject the theory as incompatible with their religious commitments, we should regard as an educational success the achievement of student knowledge and understanding of the theory and the way in which its supporting evidence makes it the best *scientific* account of the relevant biological phenomena.[3]

To foreshadow the argument to follow: in many, perhaps most, cases, belief follows naturally from understanding. When we come to understand contemporary chemistry's accounts of valence and bonding, we ordinarily come to believe that atoms combine that way; when we learn that and understand why entropy increases in closed systems, we believe that such systems become more disordered over time; when we first grasp that "every action has an equal and opposite reaction," that is, Newton's third law, we generally believe it—and so there is normally no need to distinguish belief from understanding in curricular contexts.

However, it is not always the case that belief follows understanding. For example, when students learn Newton's first law—the law of inertia, which states that objects in a state of uniform motion continue in that state of motion until influenced by an external force—many of them understand it well enough to do well on their physics exams, but when they are tested in the psychology laboratory, they often evince that they do not to believe it and instead believe that the world works as Aristotle thought, namely that objects "should go in the direction they are pushed."[4] That is, they expect moving objects to behave as if they are at rest rather than in a state of uniform motion; their expectations essentially ignore existing momentum.[5] So here—inertia—is a place where belief and understanding come apart. Evolution is of course another such place: students can come to understand it without believing it.

For students who are resistant to such belief, we will argue, knowledge and understanding are suitable results. That is, the aim of evolution education should be to enable students to *know* and *understand* the theory and its associated evidence, facts, phenomena, processes, history, and methodology. The aim needn't be to get students to *believe* it. Whether or not they actually come to believe it will be a function of many factors, both in and out of the classroom. Evolution education can be successful even though students end the course with the same nonevolutionary beliefs with which they started the course as long as they understand its precepts. At least, it is this view of the matter that we will lay out and defend here.

WHAT ARE THE AIMS OF SCIENCE EDUCATION?

Before addressing the particular case of the teaching of evolution, it is worthwhile to consider briefly the aims of science education more generally. What are we trying to accomplish in science classes? One of our aims, clearly enough, is to convey to students the *content* of our current theories. We want them to know what current scientific theory tells us about the nature of atoms, genes, molecules, space-time, and so on. In addition, we want them to know something about *how that knowledge was arrived at*: how did scientists figure out that, for instance, the genetic material is composed of DNA and space is "curved"? That is, we want students to understand how what scientists do—the method(s) they use—delivers such knowledge, and we want them to know something of the historical struggle for scientific knowledge. We also want them to have a grasp of the *epistemic status* of our theories: how does evidence support (or fail to support) particular theories? Why are we entitled to think that some theories are better supported by evidence than others?

The crucial thing in all this, we think, and fundamental to science education, is that students understand the nature and role of *reasons and evidence* in science. Central to scientific inquiry is the quest for reasons and evidence, and the point of much scientific activity is to gather and evaluate them. Experiments are designed to produce evidence that will test—confirm or disconfirm—the hypothesis under investigation. In sciences in which experimentation is for practical reasons difficult or impossible, for example in certain portions of astrophysics in which we are unable to manipulate the heavens in order to test particular hypotheses, suitable observations are made and thought experiments (as opposed to manipulative experiments), modeling, and simulations are conducted in order to produce reasons that favor or count against such hypotheses. In all of this, reasons and evidence are key.[6]

Supposing that this is right, should we think that science education aims at student belief?

SHOULD BELIEF BE A GOAL OF SCIENCE EDUCATION?

Should the teaching of evolution aim at getting students to believe the theory? In arguing for a negative answer to this question, we will address constraints properly governing the teaching of evolution to students whose religious beliefs render this belief difficult or impossible.

It is uncontroversial that science education aims at fostering student *knowledge* of particular curriculum content: theories, methods, techniques, and such. Does it follow that science education aims at fostering *belief*? As a matter of logic, the first answer must be that it does, since belief is a necessary condition of knowledge: students cannot be said to know, for example, that molecules are composed of atoms, if they don't believe it.[7] But this quick answer is problematic. We address this problem in two ways: by urging that a further aim of science education, complementary to that of fostering student knowledge, is the fostering of student *understanding*; and by clarifying the *content* of the student belief in question, that is, what specifically it is that we want science students to believe.

Consider a high-school physics class learning about the second law of thermodynamics. When we say that we want students to *know* and *understand* that entropy increases (in closed systems), what exactly do we want them to *believe*? Let us take these italicized terms one at a time.

Because justification is a condition of knowledge, *knowing* the second law of thermodynamics involves appreciating its justification: that is, why we take the law to be (approximately) true.[8] As with most important scientific content, in this case the considerations that justify the second law are diverse, involving (among other things) its derivation from higher-level physical theories, the observational and experimental evidence that confirms it, its application to real systems (for example, steam engines and other machines and processes), and so on. Knowing the second law involves more than simply writing "entropy increases" on the exam; it involves appreciating at least some of the reasons that justify it, that is, render it worthy of belief.

But of course what we've just called "appreciating" is related to *understanding*. For a student to understand the second law, she must appreciate the evidence for it—not just be able to identify on a test what the supporting evidence is, but understand why that evidence counts as evidence that supports the second law. Without such understanding, the student can't be said to know the law at all, because she has no reason for thinking it true (other than that the teacher said so).

This is not the place to attempt a serious analysis or theory of "understanding." But we can say this much, which we hope is uncontroversial. Understanding the second law requires, first, that students can identify and define the concepts involved in entropy (order, molecules, temperature, heat, and so on), and can provide rich and appropriate explanations of the interconnections among these concepts. Second, the student who understands the second law can also explain how it applies in a variety of concrete situations. Third, if he/she understands the second law, the student can appropriately apply it to previously unencountered contexts and problems. Fourth, as we've already noted, the student must appreciate at least some of the reasons why the second law is thought to be true (that render it worthy of belief), that is, she must understand why those reasons count as supporting the second law, and can appropriately evaluate the merits of those reasons.[9]

Few science educators would disagree with these goals for teaching the second law of thermodynamics. But the question concerning belief has yet to be addressed. Let us face it squarely now: if students know and understand the second law, must they *believe* it? Here we come to the central question of this chapter. There are certain propositions such that if students understand the second law, they will believe them: that the second law mentions the term *entropy*; that that term is defined in terms of increasing disorder; that the law applies only to closed systems; and so on. If students do not believe such claims, we would have no reason to think that they either know or understand the second law.

But what about the second law itself: must students believe *it*? That is, if students know and understand the second law, must they believe that *entropy increases in closed systems*? The answer here is, perhaps surprisingly, not straightforward. As we have seen, if students know and understand the second law, and if such knowledge and understanding involve appreciating its justification, then they appreciate the reasons for regarding it as worthy of belief. If they agree that those reasons render it worthy of belief, then their appreciation of the epistemic force of those reasons will lead, *ceteris paribus*, to belief. Here belief is not something requiring some extra effort or commitment; it is simply what normally results when cognitive agents appreciate the force of relevant reasons for the candidate belief in question. For example, once we appreciate the reasons for believing that touching a hot object can cause pain—in the usual case, by touching one and feeling pain—we typically believe it (and act accordingly). Similarly, once we appreciate, by working through them, the proofs of familiar mathematical claims such as the Pythagorean theorem (in other words, that $a^2 + b^2 = c^2$ for Euclidean systems) or that there is no highest prime number, we typically simply find ourselves believing those claims. In the same way, once we appreciate the reasons for believing typical scientific claims of the sort taught in school (for example, that atoms combine in particular ways, that cells divide, that smoking causes a variety of diseases, and so on), we normally believe them. In brief, *understanding typically yields belief.*[10]

Belief does not, however, *always* follow understanding; that is why the qualifier "typically" is necessary. We noted above the case of inertia as one in which belief in fact often does not follow. Science teachers will also be familiar with another type of case in which understanding does not involve belief. They often ask students to understand historically significant theories and explanations, although they have been supplanted by more recent theories. A person/student can clearly understand the principles of Ptolemaic astronomy or Lamarckian inheritance, but not believe them to be true. In yet a different type of case, a student in a class about world religions might understand the teachings of the Buddha but nevertheless not be a Buddhist (not believe in the divinity of the Buddha or believe that his teachings are true). In all such cases, belief is not an appropriate aim or expectation of public education.

Of course, the case we are worried about is the obvious one, the subject of much discussion (and hand-wringing) in the science education literature: the creationist student who can demonstrate a meaningful understanding of the tenets of evolutionary theory and the relationship of those tenets to the biochemical, ecological, paleontological, statistical, and other evidence that supports them, yet fails to believe the theory, that is, fails to believe that the theory is true. The reasons for such a response to evolution instruction can vary, but might involve a worldview that includes a deep distrust of science (and other things secular), a particular philosophical view about the relationship between science and religion, or a rejection of the epistemological presuppositions underwriting the claim that the reasons/evidence just mentioned actually constitutes good reasons for belief. In such cases, knowledge and understanding of evolutionary theory and the evidentiary support for it might not yield belief.

In cases like this, in which students master the relevant scientific content but do not believe it, how should teachers respond? Should they understand their task to be that of changing student beliefs? Should we regard their teaching as deficient if students know and understand a theory but nevertheless do not believe it? In such situations, we maintain that teachers ought *not* to strive to shape directly the content of student belief—striving to do so is the mark of the indoctrinator rather than the educator. In a crucially important sense, what students believe must be *up to them*. In such cases, we maintain that an appropriate goal is for the student to recognize the scientific status of the theory in question, that is, believe that the theory reflects the best available evidence and so affords the best current scientific account of the phenomena the theory addresses. Instruction must provide students with an understanding of the evidence related to the theory, but in the end each student must judge for himself/herself the merits of the theory's claims. Anything less amounts to a failure to treat students with respect as persons.[11]

In those cases where belief does not follow from knowledge and understanding, knowledge and understanding must, and should, suffice. In particular, if students continue to disbelieve evolutionary theory in favor of contrary religious belief, despite having knowledge and understanding of the theory, the science teacher should be content simply to point out that regardless of whatever else might be said in its favor, religious belief cannot be *scientifically* sanctioned. It is not part of the biology teacher's task to demonstrate that a religious belief is false—after all, whether science and religious belief are compatible is itself a controversial question. Neither is it the science teacher's task to judge the epistemic status of the students' religious beliefs. If a student has mastered the content of evolutionary theory and appreciates its scientific credentials, the teacher has admirably accomplished her task, whether or not the student believes the theory in the end. Indeed, the teacher may also understand but not believe, that is, not believe that evolutionary theory is true. Knowledge and understanding are sufficient goals for both student *and* teacher. To insist on student belief *in* evolution, that is, belief that it is true, is to overreach.

We maintain that in some circumstances, a person can indeed understand a theory and its supporting evidence and simultaneously refuse to believe it, and we have offered several examples of such cases above. Notice that in the case in which a student understands a theory (such as evolution) but does not believe that theory to be true, what he/she understands about the theory is not identical with what he/she does not believe. Thus, the student might understand, know, and so believe that

> *p*: *according to evolutionary theory*, speciation is a complex, lengthy process that typically takes place in contexts of geographical isolation,

but not believe that

> *q*: speciation is a complex, lengthy process that typically takes place in contexts of geographical isolation.

This case is like the one mentioned above concerning Newton's first law, in which understanding does not routinely result in belief. In that case, the student who understands the first law believes that

> *r*: *according to the first law*, objects in a state of uniform motion continue in that state of motion until influenced by an external force,

but need not believe that

> *s*: objects in a state of uniform motion continue in that state of motion until influenced by an external force.

The difference between the two cases is this. In the inertia case and similar cases, it is unusual for a student's distrust of science or view of the relationship between science and religion to interfere with her understanding of or belief in the claims understood. But in the evolution case, such interference is much more common. Thus, biology teachers may encounter many students who understand evolutionary theory but do not believe that it is true. Physics teachers also frequently encounter students who understand the law of inertia but do not believe it—but when such students don't believe it, their lack of belief is not a function of conflict with religious belief. Nevertheless, the cases are on a par: in both, the primary aim should be for students to know and understand the relevant science. If understanding is achieved, belief will typically follow, barring some specific barrier, whether psychological, religious, or some other sort. Whether or not it follows, the science teacher will have done her job well if her students acquire the relevant knowledge and understanding.

If our analysis of these several cases is correct, knowledge and understanding of specific scientific content do not require or entail belief of that same content, and it is not necessary to think of such belief as itself an aim of science education.[12] What, then, should biology teachers do when faced with students whose rejection of the theory is religiously grounded? We urge that teachers not attempt to get such students to abandon their religious beliefs, but rather to help them to understand that those beliefs are not sanctioned by science; and to help all students—including especially those who do not believe it—to understand the theory and the evidence for it, and that that evidence renders the theory the best current *scientific* account of the relevant phenomena. If the disbelieving student has a full and meaningful understanding of evolution's tenets and scientific credentials, understands that her rejection of evolution is not itself scientifically based, and believes that evolution affords the best current scientific account available given the available evidence, then she has achieved our most important legitimate educational goals—goals that are all too rarely achieved.

We began this chapter by asking what a science teacher should do when confronted with a student who does not believe the science content he/she has learned. Our answer is: strive for student knowledge and understanding, and let belief fall where it may. Typically, belief will follow such knowledge and understanding; when it does not, that disconnect is usually the result of strongly held convictions that are themselves extrascientific and so beyond the bounds of science and science education to resolve. Students who raise such questions in the classroom present the teacher with an opportune setting for teaching about the nature of science, and in particular the role of reasons and evidence in science. Science teachers should grasp this opportunity to help students carefully distinguish knowing, believing, and understanding, and, in particular, to stress the empirical nature of scientific inquiry and the limits of such inquiry.[13]

We turn next to a dimension of the creationism/evolution controversy made much of in our earlier historical chapters: the controversy concerns more than science and has deep cultural roots. Does regarding the controversy as a clash of cultures, rather than simply rival visions of science, suggest alternative educational proposals aside from those we've advanced thus far? In the next chapter, we treat the conflict in cultural terms and explore its educational implications when so conceived. We urge that the conflict is indeed appropriately seen in cultural terms, and that seeing it in that way enables us to put forward curricular and policy recommendations that recognize the legitimate claims of both evolution supporters and evolution opponents and gives both sides their due.

CHAPTER 8

A Question of Culture?

Though the terms of the debate have changed drastically since the 1920s, the teaching of evolution in public high-school biology classes remains as controversial and contested today as it ever has been throughout the long history of the conflict. Despite the fact that scientists virtually unanimously endorse contemporary evolutionary theory (henceforth ET) and reject the creationist/ID alternatives as scientifically deficient and so unworthy of inclusion in the biology curriculum, evolution opponents continue unabated in their efforts to both inhibit the teaching of evolution and increase the teaching of creationism/ID. On the scientific merits, the issue should be long settled: biological scientists teach, develop, and utilize ET in their research and have achieved stunning success that would not have been possible without the insights provided by ET.[1] On the other hand, as has often been pointed out, there is no established body of creationist/ID-inspired scientific discoveries or even an identifiable research agenda. Given the success of ET and the failure of ID, why is this battle still being fought?

A possible explanation is that the dispute is not a scientific one at all, but rather a thinly veiled conflict of cultures: those who reject ET in favor of ID are moved primarily not by scientific disagreement but by competing cultural visions and values. They view ET not (just) as bad science but as essentially connected to values they abhor (such as secularism) and cultural visions they find empty, unsatisfying, sometimes terrifying, but in any case devoid of hope, value, and purpose.

In this chapter, we consider the controversy as a cultural conflict. What is involved in viewing it as essentially a cultural clash rather than a scientific controversy? If we agree that the controversy is best seen in cultural terms, what follows concerning the teaching of ET in public schools? In particular, does viewing the advocates of ID as representatives of an aggrieved cultural minority force a reconsideration of the aims of science education? Are these aims best seen as culturally relative or culture specific? Does "aggrieved cultural minority" status provide a good reason for either limiting the teaching of evolution or expanding the teaching of creationism/ID in the science classroom? These are the questions we take up here. We begin with a brief look at *multiculturalism* and the educational demands it might make.

MULTICULTURALISM VERSUS UNIVERSALISM: A FALSE DICHOTOMY?

Multiculturalism is a widely used term. It is understood in different ways by different authors. We cannot review here its many meanings extant in the literature; rather, we will stipulate what we mean by it in what follows. As we use the term here, *multiculturalism* refers to views, policies, and practices that advocate the celebration of cultural differences; insist on the just, respectful treatment of members of all cultures, especially those that have historically been the victims of domination and oppression; and urge full recognition of the integrity of historically marginalized cultures.[2]

Many philosophers and education scholars[3] urge that educational efforts reflect a basic and fundamental commitment to multiculturalist principles and practices. In particular, they urge that educational efforts abandon all "universalist" pretenses and acknowledge the cultural specificity of all educational content and the centrality of cultural concerns to virtually all matters of educational policy. This multiculturalist view of education rests on an epistemologically contentious view of multiculturalism itself, according to which allegedly "universalistic" ideals—for example, moral, social, epistemic, or intellectual ideals such as justice, equality, and rationality—are not, and indeed cannot be, universal. Rather, they are merely *local* ideals: acknowledged and embraced in particular cultures and, by virtue of those cultures' hegemonic domination over other cultures, wrongly imposed on those dominated others and falsely proclaimed as universal.

Is it correct that "any insistence on the universalism of values must be no more than the projected imposition of local values . . . universalized"?[4] Are all such "universalizations" of values and ideals nothing more than the hegemonic imposition of the values and ideals of another culture on dominated others? Are all allegedly universal values mistakenly so proclaimed? Are they all *merely* local? In particular, is the value of multiculturalism itself merely local?

We think that the universal/local dichotomy presupposed by many leading figures in the multicultural educational literature fails. Indeed, as we argue next, accounts of multiculturalism that preclude the very possibility of universal ideals are incoherent; and in light of the standard justification of it on moral grounds offered by its leading philosophical and educational proponents, multiculturalism must itself be regarded as a universal social/political/educational ideal.[5] We endorse multiculturalism, as we have defined it, but deny that we must therefore reject universal values or ideals.

In claiming that some particular value or ideal is "universal"—like those of justice, equality, human dignity, rationality, or that of multiculturalism itself—the universalist need not, and ought not, reject the completely compatible claim that these ideals are also "local" in the sense that they have been formulated and advanced in particular historical/cultural locations and recognized and endorsed only by some, but not all, cultures. In holding these ideals to be universal or transcultural, the universalist is completely free to acknowledge their locality/particularity. Many authors in this literature presume that "local" and "universal" are contradictories and so that "local" entails "nonuniversal," but this is simply an error. It is of course correct that all proclamations of universal principle emanate from and are championed in particular locations. It does not follow from this, though, that such principles have no legitimacy or force beyond the bounds of the locations from within which they are proclaimed. All principles, values, and ideals—indeed, all beliefs, theories, and judgments more generally—are conceived and embraced (if at all) only in particular cultural locales; whether or not any of them enjoy legitimacy beyond the bounds of those locales is an independent matter. When such legitimacy extends across all local boundaries, they are both local *and* universal: local in formulation and perhaps in acceptance, but universal in applicability. This dichotomy is a false one—and so, one cannot rightly reject universality on the grounds that all ideals are local, either in origin or in current acknowledgement and acceptance. As Israel Scheffler tellingly puts the point: "I have always supposed that the universal and the particular are compatible, that grounding in a particular historical and cultural matrix is inevitable and could not conceivably be in conflict with universal principles."[6]

It is often thought that universality must be rejected because ideals and values, in order to be "universal" or "transcultural," must be grounded in some impossibly neutral, "perspectiveless" perspective or God's-eye point of view.

The thought is that since such a "view from nowhere"[7] is impossible (for creatures like us, at any rate), so are universal values and ideals.

We agree that so understood, there can be no such universal values, principles, or ideals. But we need not and should not understand the term in this way. Let us grant that there are no universal or transcultural values in the sense that they are grounded in, or derived from, a perspective outside history and culture, for there simply is no such "Godly" perspective available to us. Nevertheless, in the more modest sense noted above—according to which a principle, value, or ideal is universal insofar as it has application across all local cultural boundaries—Godly perspectives or perspectives outside history or culture are not required in order for universal status to be achieved. Achieving it is a matter not of impossible perspective but of applicability across local boundaries. To say that values and ideals are universal is not to say they are from God or from Nowhere but rather that they transcend individual cultures—they are legitimately applicable and have force not only beyond the bounds of the particular cultures in which they are acknowledged but also beyond all such cultural boundaries. In this they are like laws of nature, mathematical truths, and moral truths such as that of multiculturalism itself: "$E = mc^2$," "Entropy never decreases in closed systems," "There is no highest prime number," "It is wrong to torture small children for the fun of it," and "All people, including members of oppressed marginalized cultural groups, should be treated with respect" are true in and applicable to the inhabitants of all cultural locations—even those in which the relevant concepts are not in play and belief in them uncommon or altogether absent. They were not given to us by God, they did not come from Nowhere, they were first articulated in particular cultural locations. Nevertheless, their truth and applicability transcend all local boundaries; they are both local *and* universal.

Moreover, it is worth noting that reasoning from the impossibility of attaining a neutral perspective to the impossibility of universal values or ideals itself presupposes a sort of argumentative universality: namely, that the premise entails the conclusion. The opponent of universalism who advances this argument takes this entailment relation to hold universally—not only locally in some cultures and not others but also across all such cultural boundaries. In this way, any form of multiculturalism that insists on a sharp universal/local dichotomy, embraces the latter, and for that reason rejects the former, itself presupposes the very sort of universalism it seeks to deny. In rejecting the dichotomy, we make possible the embrace of a logically consistent form of multiculturalism, one that understands multiculturalism as *itself* a universal ideal that urges the celebration of cultural differences; the just, respectful treatment of members of all cultures, especially those which have historically been the victims of domination and oppression; and the integrity of historically marginalized cultures.[8]

MULTICULTURALISM, UNIVERSALISM, AND "ETHNIC SCIENCE"

What are the curricular ramifications of the "local-only" version of multiculturalism? In particular, does it follow from that view that the embrace of multiculturalism in particular curriculum domains requires the rejection of universalistic conceptions in/of those domains?

The question is particularly pressing for science education. Perhaps surprisingly, many science educators have in recent years argued that science education should abandon its universalistic pretenses and acknowledge the cultural specificity of all scientific theories, including ET. Should science educators conceive of science as *universal* or *culturally transcendent*, or should we rather think of science as *culturally specific* such that different cultural groups have their own sciences or, more modestly, that what should be taught in public-school science classes depends on the cultural makeup of students or their community?

This question gains force if we think that advocates of creationism/ID are best understood as making a point about their culture and its values rather than the standing of ET and ID conceived as culturally transcendent theories of the biological world.[9] Should science curricula treat Western Modern Science (WMS)[10]—that is, the physics, chemistry, biology, and other scientific disciplines taught in leading universities in North America, Western Europe, Asia, and elsewhere—as just one local alternative among others? Should science teachers teach local "ethnic sciences" in their classrooms? If so, should they teach them as the epistemic equivalents of the allegedly erroneously proclaimed universal alternative? Prominent science education scholars have advocated all these suggestions, concerning both multicultural education in general and multicultural science education (MSE) in particular.[11] If not, on what basis can/should WMS be privileged?

There are many examples of "ethnic" or "indigenous science"—Maori, Islamic, African, Alaskan native, and so on—discussed in the science education literature. Some discussions point out the impressive observational discoveries and ecological triumphs of local groups, others the very different metaphysical bases of their "explanations" in spiritual, magical, or animistic terms. It should be granted that "ethnic sciences" have indeed successfully revealed important knowledge of natural systems and environments. But this knowledge is essentially limited to naturalistic observation. It does not generally go further to construct testable, predictive, and explanatory theories concerning such observation. For that reason, the understanding of nature it provides is *less adequate, qua understanding the natural world*, than that of WMS. That is, ethnic sciences have not been as successful as WMS in developing the genuinely testable, predictive, and explanatory sort of scientific theories prized by universalists and required by the criteria of scientific quality canvassed in chapter 5.[12]

All the parties to this dispute think of scientific theories and practices as constructed by historically and culturally located people and that science provides only one among a wide range of ways in which humans have attempted to come to grips with the natural world. But those who embrace the universal character of science also believe that from among that range of ways of understanding the world, WMS is the most successful way of understanding it that we have, insofar as success is measured in terms of the production of the testable, predictive, and explanatory theories that are the hallmark of high-quality science. This in turn provides the answer to our earlier question concerning the privileging of WMS. In what sense, if any, is WMS rightly thought to be scientifically superior to its ethnic alternatives? The answer is clear: while some "ethnic sciences" have indeed recorded some impressive achievements, especially involving subtle observation and ecological management, those achievements do not do as well in terms of the criteria laid out here and in earlier chapters. WMS strives to develop theories that are testable and explanatorily powerful, make accurate and sometimes novel predictions, afford understanding of underlying causal mechanisms and of causally important unobservable entities, and are open to critical assessment. Its most important theories satisfy these criteria to an impressive degree. "Ethnic sciences" satisfy them less well (if at all) and for that reason are less worthy of inclusion in the science curriculum.

This superiority is limited: WMS is not *culturally* superior in any way. Its superiority is strictly scientific, and lies in its production of theories that satisfy these criteria to an impressive degree. This is a restricted but nevertheless crucially important superiority; it is what entitles WMS to its place at the center of science education. We can and should happily acknowledge the many successes that "ethnic sciences" have enjoyed, and we should not hesitate to incorporate them into the science curriculum where appropriate. Nevertheless, we should also not hesitate to say that WMS is superior insofar as it has produced a deeper understanding of the natural world by producing testable, predictive, and explanatory theories concerning the underlying causal mechanisms and statistical regularities that account for the features of the world we observe. Universalists should not deny that many of the successes of "local," "ethnic sciences" are important or useful. The many examples of local ecological knowledge detailed by Snively and Corsiglia, El-Hani and Bandeira,[13] and others are especially impressive. Nevertheless, because these "ethnic sciences" do not in general concern themselves with prediction, explanation, experimentation, theory construction, or the discovery of underlying causal mechanisms, they by and large fail to satisfy the criteria of good science by which WMS aspires to measure itself.

Are these criteria of scientific adequacy—testability, explanatory power, predictive power, and the like—themselves morally problematic? Does their assumption in science education contexts amount to a morally repugnant form of cultural imperialism, domination, or hegemony?

The short answer is no: respecting students from minority cultures does not require regarding their "ethnic science" as the epistemic equal of WMS. Respecting them and their cultures involves taking them seriously, not rubber-stamping the quality or legitimacy of their cultures' understanding of the natural world, and includes making clear how and why they fail to meet the criteria of WMS (when they do). This is a crucially important part of students' science education. The aim is not to get students to reject their culture; it is rather to impart a serious understanding of WMS.[14]

Of course, such teaching, as all teaching, must treat students respectfully. As many defenders of MSE forcefully point out, teaching WMS to students from minority cultures—especially those that propound supernatural or other culturally specific understandings of the natural world—in insensitive or demeaning ways can have a detrimental effect on students, leaving them alienated from their own culture and feeling ignorant, intimidated, inferior, and worse. Such a deleterious result is surely not what we want for students and would suggest a gross dereliction of our duties as educators, and advocates of MSE are certainly right to decry this sort of pedagogical abuse. Science educators must endeavor to teach all students, including those from minority cultures, sensitively and with respect. At the same time, we should acknowledge that such respect does not require giving up the criteria of WMS or regarding "ethnic sciences" as being on an epistemic par with it.[15]

To sum up this section: if we want science students to understand our deepest extant accounts of the natural world, which take the form of testable and confirmed theories that enjoy both explanatory and predictive power, and to appreciate both the power and the limitations of the methods that scientists have utilized in order to yield that understanding and the tentativeness and fallibility of all such theorizing and understanding, there is no serious alternative to the privileging of WMS in the science curriculum when seeking to understand the natural world. We should not hesitate to discuss other, "local" or "ethnic" alternative approaches to or understandings of the natural world, nor should we shy away from considering challenges to WMS made from the perspective of those alternative approaches. At the same time, we should help all science students understand the features of WMS that enable it to achieve the limited sort of superiority just noted.[16]

EVOLUTION VERSUS CREATIONISM/ INTELLIGENT DESIGN AS A CULTURAL (RATHER THAN SCIENTIFIC) CONFLICT

We have argued thus far in this chapter that the local/universal dichotomy is a false one, and that science education should acknowledge both the local and the universal dimensions of science. We have also argued that although all science students, and in particular those raised in minority cultures, must be treated and taught respectfully, this obligation does not challenge or upend the educational directive to teach students the strengths of WMS and the limited but important superiority it enjoys. How does all this relate to the evolution versus creationism/ID conflict?

The battle over the teaching of evolution is often viewed as a clash of cultures. Is it accurate or instructive to view the advocates and opponents of the teaching of evolution as members of distinct, incompatible cultures? It is difficult to define *culture* or individuate cultures in uncontroversial ways. We will not offer an account of cultural difference here, contenting ourselves with understanding *cultural difference* as a matter of degree involving differences in basic beliefs, values, and practices. So understood, it is tempting to conceive of the partisans in this conflict as inhabiting different cultures. Even if the two sides are rightly thought to be members of competing, incompatible cultures, however, we argue next that nothing much follows from this assumption concerning the teaching either of evolution or of competing accounts of creation. Whatever cultures students call their own, their science education should enable them to understand ET and understand that that theory is the best *scientific* contender in the domain.[17]

What is involved in conceiving of the evolution versus creationism/ID conflict in cultural rather than scientific terms? Supposing that there is a legitimate way of privileging WMS, as we have argued, does it follow that this settles the question of biology curriculum content in favor of evolution and against ID? It would be convenient for evolution supporters if this were so, but thinking that such privilege does so settle the question might be thought to ignore the specifically cultural nature of the conflict, if it is indeed rightly seen as a cultural (rather than a scientific) conflict. After all, in the United States education is often declared to be, and in fact largely is, a local matter—"local control" is a rallying cry in the political arena—and parents are frequently thought to have the right to determine the curriculum content of their children's schools and, more dramatically, to impose their religious beliefs on local children in ways that prevent students from subjecting those beliefs to critical scrutiny.[18] More broadly, multiculturalists frequently proclaim the imperative of "respecting and maintaining" cultural diversity. Should we then forego the teaching of ET and allow the teaching of creationism/ID in the high-school biology class, not because it is good science but because it represents or constitutes an important cultural value that deserves protection on multiculturalist grounds?

We think this question must be answered in the negative: however important it is to support cultural values, it is not so important as to sacrifice otherwise unobjectionable curriculum content, or include objectionable content and treat it as if it were unobjectionable.[19] This is one of the lessons of the previous section. We argued in chapter 7 that understanding does not always or necessarily result in belief, and that students who understand the science of evolution but reject it for religious or cultural reasons are not thereby science-educational failures. But science education *has* failed those students whose religious or cultural commitments prohibit them from *understanding* science. Students whose science education has not helped them understand the phenomena of evolution and the evidence and theory that make it the best scientific understanding of its domain—or worse, whose science education has frustrated their understanding of the phenomena and our best scientific accounts of them—have been harmed by that ineffectual and/or obstructionist education.

There is nothing special about evolution in this regard. Any content deemed worthy of curricular inclusion is so deemed for a variety of reasons, both intrinsic and instrumental. We want students to understand what we teach in math, science, history, literature, and other classes because of its intrinsic value, but also because we think it will help them to develop critical thinking skills and dispositions, analytical and communication abilities, and other educational goods that will enable them to live satisfying and productive lives. This is true of education across the curriculum; it is equally true of science education in general and evolution education in particular.[20]

It is worth noting that the same criticism can be made of ineffectual or obstructionist science education that provides students a "rhetoric of conclusions"[21] concerning evolution but no real understanding of it or its evidential credentials. Evolution education can be problematic in some of the same ways that creationist/ID education can, and students can come away from evolution education with firm, even dogmatic, belief in its truth but with no genuine understanding of either it or the evidence that supports it. Such a pro-evolution education, which results in students whose commitment to and belief in ET is dogmatic but lacks understanding, is as problematic as the anti-evolution/pro-creationism/ID alternative. Genuine understanding of the biological world—a fundamental aim of biology education—is absent in both cases.[22] As Mary Midgley engagingly puts the point, we should teach students "in a way that will make it quite impossible for them—whatever views they started with—to come out either as bigoted fundamentalists or bigoted atheistical pseudo-Darwinists."[23] Moreover, the aim of helping students to achieve understanding is not itself culturally bound. Without this aim, the activity in question, whatever its other merits, would hardly qualify as *educational*.

Even if the conflict is indeed best seen as cultural rather than scientific, the presumed imperatives of fostering cultural diversity and respecting cultural values must be reconciled with the need to offer biology students the best extant scientific account of the domain. We have already argued that teaching creationism/ID as the scientific equal of evolution fails to meet the criteria of good science identified earlier. Teaching the best extant scientific account of the domain takes priority over students' and parents' claims to cultural accommodation. "Cultural values" justifications of curriculum content may have some legitimate force, but they also face inherent limits imposed by the broader aims of education. For that reason, parents and other advocates of cultural accommodation cannot legitimately keep their students from learning important curriculum content simply because it is thought to conflict with their deeply held cultural convictions. To do so would be to give up on education itself. As we have already noted, educators must treat all students, including and especially those from marginalized cultures, respectfully.[24] This is true of students from "creationist cultures" as well, of course. But if respectful treatment of such students involves *educating* them, there is no legitimate alternative to endeavoring to foster in them an understanding of ET, its evidential credentials, and its status as the best *scientific* account of the biological world.

If so, legitimate multicultural imperatives to respect and maintain cultural difference must be balanced against our obligation to expose students to the best disciplinary understanding of the relevant curriculum domain. This means that in biology class, while we needn't and shouldn't endeavor to dissuade students of their cultural values and understandings, we must do our best to enable them to understand the reasons that establish ET as the best scientific account of the biological world. We needn't endeavor to get students to *believe* it, but rather to *understand* its scientific credentials.

We conclude that (1) multicultural science education is compatible with a universalistic conception of science; (2) multiculturalism requires the teaching—to all students of all cultures—of that universalistic conception; and (3) however well intentioned, it is both a moral and an epistemic mistake to teach science as if alternative "ethnic sciences" are, by virtue of their places in the cultures that embrace them, of equal epistemic status. Even if the evolution versus creationism/ID controversy is indeed most instructively seen as a cultural rather than a scientific matter, as we have suggested, that construal of the controversy does not provide a good reason to think that creationism/ID deserves equal treatment with evolution in the science curriculum (though it may well deserve a place in other subject classes, for example in history or social studies classes, so long as it is not presented as scientifically credible). In the end, whether or not this dispute is properly conceived as a matter of competing cultures, it is not resolved in this way. Criteria of scientific quality and the goal of fostering student understanding trump the preservation of culturally specific beliefs in public science education.

CONCLUSION

Evolution as Education

What is the evolution/creationism controversy all about? Seen from one perspective, it might be a problem of "dead science."[1] From another, it might boil down to the political power of teachers.[2] From a third, it might be best understood through the lens of "social movement theory."[3] All these insights help us understand the nature of the long dispute over the teaching of evolution. But none of them put education squarely at the center of their view. This volume has combined historical with philosophical approaches to these controversies. In each case, we think that a sensible emphasis on the evolution/creationism controversy as a matter of educational dissent will suggest new approaches and new perspectives.

Comparing the history of evolution education to other controversial issues in American education, for instance, we see that evolution and creationism haven't been trapped in any sort of deadlock, at least not more so than any other stubborn cultural disagreement. During the 1920s, as the historical chapters in this volume have demonstrated, those who wanted to promote evolution education wondered—with good reason—if evolution would ever be allowed in public schools throughout much of the country. By the twenty-first century, in stark contrast, the situation has been dramatically transformed. Now, evolution's place in public schools is more secure. The problem instead is that so many Americans seem to welcome nonevolutionary explanations into their children's science classes. Evolution education is often watered down or intellectually mutilated, but it is no longer legally banned—at least in the United States. The notion that evolution education has stalled or deadlocked is simply untrue. Only if we speak in the broadest terms is it historically accurate to say that education policy has been trapped in the same controversy for almost a century. And this could be said for every issue on which Americans disagree. Race and racism, for example, remain stubbornly controversial as parts of America's public schools. But does that mean there has been no substantial change in racial attitudes in America's schools? That no progress has been made?

True, evolution has remained controversial, but the positions of the protagonists in the controversy have largely reversed themselves. In the early twentieth century, biologists such as Edwin Conklin pleaded desperately with Americans to allow evolution into classrooms. Today, there is no such existential threat to the teaching of evolution in America's public schools. To imply that the issue has ground to a deadlock suggests that radical action is necessary. An accurate picture of the history of this controversy, however, suggests different policies.

There have been enormous and radical transformations in the realities of evolution education in the past century. More American schoolchildren spend more time learning evolution than ever before. Is today's evolution education perfect? Far from it. But there is a significant difference between a policy that calls for deepening and broadening current educational efforts and one that calls for a radical transformation. There is no reason for complacency, but taking the long view, evolution education is winning.

There are additional policy implications to viewing the history of the controversy over evolution education as merely one aspect of the broader history of cultural dissent in education. Though the fact might make activists from both sides uncomfortable, we maintain that the best way to understand creationism is as part of America's long tradition of religious dissent and cultural pluralism in education. When we understand creationist parents, communities, and students as religious and cultural dissenters, it may generate more productive approaches than if we view them as somehow ignorant, or perhaps innocent, of evolutionary science. This simple historical truth has been complicated by the fact that both mainstream scientists and creationist dissenters have tended to insist that the central issues of the debate concerned science and the nature of science. Time and again, each side has managed to talk past the other by asserting different conceptions of science.

If instead we view creationism as part of America's storied legacy of religious dissent and cultural pluralism in schools, we can see new policy approaches to bridging the evolution/creationism divide. America has a checkered history of accommodating cultural minorities in its public schools. Throughout most of the history of public schooling, public-school leaders have assumed too much about the nature of America's cultural consensus. In the nineteenth century, notably, Catholic dissenters often felt shunted aside by a Protestant school establishment blind to its own prejudice.[4] In the twentieth century, nonbelievers felt pressured to accept theism by consensus as part and parcel of their children's public-school experience.[5] In each case, school policies brushed off the objections of dissenters as unreasonable or outside the pale of mainstream American society.

What difference would it make if we viewed creationists as cultural dissenters with the rights and responsibilities due to any minority group? One thing it would not mean is that the children of creationist families would somehow get a pass from learning about evolutionary science in public-school science classes. Nor would it mean that public schools would need to teach creationism as science to avoid trampling the civil rights of creationist students. Teachers, for example, don't read the Catholic catechism in public-school classes. Nor do atheist students get a special class that trounces the arguments for a loving God. Rather, what we can learn from America's ugly history of trampling dissent and minority rights in public schools is that public schools must create a community broad enough to include dissenters on equal terms. Creationist students and families should be treated with the same respect to which all dissenters and minority groups are entitled. They must not be treated as if their dissent implies a deficit of some kind, especially a deficient religion. Such judgments must remain wholly beyond the authority of teachers and administrators.

This isn't special treatment for creationists. Rather, as the philosophical chapters in this volume have demonstrated, this simply means treating creationist students as we treat students of every minority or dissenting background. Treating them in this way would imply nothing concerning the content of biology classes. As we have seen in chapter 8, treating such students with respect doesn't require teaching biology as if creationism or ID were scientifically legitimate alternatives to evolution. Rather, it requires sustained attention to the criteria marking high-quality science and the ways in which ET, creationism, and ID—and perhaps other "ethnic sciences" as well—satisfy or fail to satisfy those criteria. Doing so respects students by taking their dissenting views sufficiently seriously so as to subject them to critical scrutiny and also by taking seriously our obligation to foster their understanding of high-quality science, including their coming to see that ET offers the best scientific understanding of the biological world. Divergent cultural backgrounds must be respected, but that doesn't mean that their culturally specific beliefs should supplant the findings of mainstream science.

Viewing creationism as a question of education for a dissenting group also offers new big-picture policy outlines. First, if we see the long-range outcome of the controversy to be moving toward more and better evolution education, we should continue with current practices. Textbooks, standards, and teacher training should all emphasize authentic evolutionary science as the best available science. We must not allow dissenting cultural beliefs to determine public-school science curricula. Efforts to improve science education, often taken at significant cost by activists in the science education community, must continue. But there are other less commonly promoted policies that will also help. If we understand creationism, historically and philosophically, as a case of minority belief, a form of religious or cultural dissent, then we must also ensure that science *teacher* education focuses on more than just biology and geology. To make better schoolteachers, we must teach them about the history and culture of creationism. Just as teachers of any ethnic minority students will teach better if they can connect culturally with their students,[6] so science teachers of religiously dissenting students will teach better if they know a thing or two about the reasons for student dissent.

Perhaps our most important recommendation is the replacement of belief with that of knowledge and understanding as fundamental aims of science education. Among scholars and science educators, the notion of emphasizing understanding over belief has gained numerous influential adherents, but it is still a matter of some controversy.[7] If some think the aim of biology instruction is to get students to believe ET—that is, believe it to be true—it is no surprise that dissenters, both students and parents, will object. In our view, their objection can be quite legitimate. For one thing, fostering belief in this way jeopardizes the satisfaction of the fundamental educational aim of fostering student *autonomy*.[8] For another, whether justified or not, it is not surprising that parents and students alike often place great value on the maintenance and continuation of central cultural and religious beliefs, and evolution education can threaten these. But it is also clear that understanding of contemporary science is deeply important for students, in terms of both their general understanding of the world and their options for adult life. Opting for the aims of knowledge and understanding rather than belief allows science educators to deliver the best of both worlds. Teachers must seek to foster knowledge and understanding of science, including evolution. Teachers must share with students the evidence that makes evolution the only genuine game in (science) town. But teachers must also honor student autonomy, and acknowledge the legitimacy of the deep interests of some students in cultural identity, continuity, and community. Teachers must also enable students to take full advantage of the fullest range of life options possible, and some of those options will depend on an authentic knowledge and understanding of ET.

The model of understanding-not-belief proposed in chapter 7 might best be illustrated by a couple of additional examples. Consider, first, the case of an applicant who was denied admission to a biology education graduate program because of his opposition to evolution, even though the student appeared to be otherwise highly qualified;[9] or the prospect of a similar applicant being denied admission to medical school.[10] Such cases are in important respects unlike the ones we have been considering, since they involve graduate training rather than public K–12 education. Nevertheless, our model might shed some light. Since these cases involve professional training, it would, we think, be appropriate for the programs in question to reject such applicants if their opposition to evolution threatened to undermine their future professional performance. This might be the case if, for example, the medical school applicant eventually became a medical researcher, or if the goal of the graduate program applicant was to use his PhD credentials to work to undermine public-school evolution education. But these are hardly inevitable results: the prospective physician might rather become a heart surgeon, ophthalmologist, or general practitioner, in which case her knowledge and understanding of evolution might well be sufficient for her success, despite her lack of belief.[11] The prospective science educator might similarly be enabled by his knowledge and understanding of evolution to be a perfectly competent science educator, effectively contributing to educational research and helping future science teachers to hone their skills, despite his lack of belief. In other words, whether or not lack of belief, coupled with knowledge and understanding of ET and its supporting evidence, is a disqualifier depends on the specifics of the case. Lack of belief should not be an automatic disqualifier or reason for rejection in such cases: sometimes it will be; sometimes not.

On the other hand, respecting the rights of dissenting groups does not mean schools should devolve into educational free-for-alls. Lawmakers in New Hampshire recently made an egregious overreach in this area. In 2012, the state legislature passed a bill that allows parents to opt out of any curricular material they consider objectionable.[12] The goal was to protect dissenting families from ideological imposition by insensitive teachers and school boards. In such a case, however, the Granite State lawmakers ignored a central purpose of public education. Schools must require students to engage with uncomfortable ideas. Curricula chosen through an open, informed, and democratic process must not be subject to individual objections on a case-by-case basis. Not only is such a system impractical, it threatens to erode the purpose of public education. If dissenting students can simply opt out, they might be able to ignore fundamental truths that every educated person should know.

In the case of evolution education, these cases demonstrate the purposes and limitations of an understanding-not-belief approach. If a student can demonstrate that he or she understands central ideas, such as a medical student who understands ET, then his or her beliefs about those ideas might be irrelevant. Public schools, in particular, must not impose on them. But this doesn't mean that public schools must not insist that students learn foundational ideas. Students and their parents should not be allowed to pick and choose between truths; if they are, they may well select only what is culturally palatable to them. Rather, schools have a responsibility to be sure that all students have, to the best of their abilities, the best knowledge and understanding available. In the case of evolution education, that means that all students must understand ET, even if they don't believe it.

Many if not most curricular areas aim to foster knowledge and understanding rather than belief. Literature courses don't aim at getting their students to believe this or that about Shakespeare, Molière, or Morrison; rather, they strive to impart to students sufficient knowledge and understanding of the texts as well as the ability to interpret and theorize so that students can form their own beliefs about the meaning and value of those texts. History courses likewise don't seek to impart any particular beliefs about this or that historical event, period, or figure, but rather to equip students with sufficient knowledge, understanding, and analytical ability so that they can competently come to their own conclusions. Similar remarks can be rightly made concerning courses in the other humanities and social sciences, and, we think, science courses as well, especially biology courses in which evolution plays a central role.

Throughout education, we should be aiming at fostering student knowledge, understanding, and critical, analytical ability and content ourselves with letting belief fall where it may. It is perhaps worth repeating the point made in chapter 7, that belief *typically* follows knowledge and understanding, and will usually not be problematic educationally. In the cases in which it is problematic, such as the case of evolution education, we urge educators not to overreach. Science educators must be trained to pursue the goal of teaching all their students to understand evolution and not to press toward the chimerical goal of demanding that they believe it. This goal of student understanding of evolutionary concepts and content should continue to form the backbone of curricular materials, standards, and evaluation. Demanding that students from creationist families change their religious beliefs is beyond the scope of public schools' mission. After all, do English teachers demand that students agree with their interpretation of Lady Macbeth's character? Or rather, do they insist that students learn to speak intelligently and cogently about ambition and guilt along with plot and imagery? Similarly, should students who speak a regional, racial, or ethnic dialect be told that their background proves their ignorance about Shakespeare's English? Or instead, should teachers work hard to understand the backgrounds of such students, even while pushing them to make sense of the witches on the moor?

Certainly, there is plenty in these policy prescriptions for both sides to dislike. Creationists will likely object to our insistence that every student in public school must learn evolutionary science and not supposed alternatives to it in their biology classes. They may protest that all students need to be taught to think critically about evolution by learning about dissenting theories. On the other side, science pundits may insist that limiting our goal merely to student understanding and not belief waters down the content of science classes. Such folks may view our suggestion to educate science teachers about the cultural background of creationist students as truckling to the demands of a loud-mouthed and self-serving minority. We hope that the fact that both sides will predictably object to our attempt to reconceive the controversy in a way that gives both sides their due indicates something promising about our proposed approach.

Evolution will not likely cease to be culturally controversial, at least in the foreseeable future. We believe, however, that understanding the history of the controversy and the philosophical issues involved will allow us to move toward both a more complete understanding of it and a more sustainable educational policy toward it. Too often, scholars and pundits have framed the evolution/creationism controversy as primarily about the nature of science or religion or politics. By viewing the bitter evolution/creationism instruction controversy as a cultural matter, we can see it in a clearer light. That clearer light will, we hope, enable us to move forward in ways that enhance the education of all our students while simultaneously treating them all with the respect to which they are entitled.

NOTES

INTRODUCTION

1. Yudhijit Bhattacharjee, "Evolution, Big Bang Polls Omitted from NSF Report," *Science-Insider*, April 8, 2010, http://news.sciencemag.org/2010/04/evolution-big-bang-polls-omitted-nsf-report (accessed March 23, 2015).

2. Controversies over creationism and evolution are certainly not limited to the United States. As Ronald L. Numbers and other historians have demonstrated, modern creationism is a global phenomenon. See Numbers, *The Creationists: From Scientific Creationism to Intelligent Design*, exp. ed. (Cambridge, MA: Harvard University Press, 2006); see also Abraham C. Flipse, "The Origins of Creationism in the Netherlands: The Evolution Debate among Twentieth-Century Dutch Neo-Calvinists," *Church History* 81 (March 2012): 104–47, and "Creationist Legislation in Brazil," *National Center for Science Education Blog*, November 21, 2014, http://ncse.com/news/2014/11/creationist-legislation-brazil-0016007 (accessed March 23, 2015).

3. The terms *believe* and *accept* are used inconsistently, and often interchangeably, in the science education literature. As philosophers use them, *believe* means (roughly) "take to be true," and is involuntary and needn't be relied on in practical reasoning, while *acceptance* is voluntary and is so relied on. For explicit definitions, examples, and extensive discussion, see Mike U. Smith and Harvey Siegel, "On the Relationship between Belief and Acceptance as Goals of Evolution Education: Ten Years Later," *Cultural Studies of Science Education* 2015 (forthcoming).

4. This tradition could with justice be understood as one of *cultural* dissent: creationists and evolution opponents (as we shall call them in what follows) are motivated not solely by religious concerns but by broader cultural ones as well. But this book is not the place to investigate the complex relationships between religion and culture. We acknowledge this complexity, but will use both "religious dissent" and "cultural dissent" in what follows as dictated by context.

5. See, e.g., Philip Kitcher, *Living with Darwin: Evolution, Design, and the Future of Faith* (New York: Oxford University Press, 2007).

6. See, e.g., Michael Berkman and Eric Plutzer, *Evolution, Creationism, and the Battle to Control America's Classrooms* (New York: Cambridge University Press, 2010).

7. Michael Lienesch, *In the Beginning: Fundamentalism, the Scopes Trial, and the Making of the Antievolution Movement* (Chapel Hill: University of North Carolina Press, 2007).

8. To make the situation even more complicated, terms such as *evolution* themselves have an infinitely complicated bundle of meanings, depending on who uses them and how. *Evolution* is used to refer both to unguided, wholly material development and to divinely guided changes in species. The terms for the various sorts of evolution can become cumbersome, with *theistic* evolution contending with *materialistic* evolution, *evolutionary creationism* contending with *organic* evolution. In this book, we will use *evolution* to refer to the changing and development of species without any divine guidance or supernatural interference. Only when it is necessary for clarity will we distinguish between various sorts of divine or nondivine evolution.

9. Kitcher, *Living with Darwin*, 8.

10. Glenn Branch, "Understanding Creationism after Kitzmiller," *BioScience* 57, no. 3 (2007): 278–84.

11. Numbers, *Creationists*, 7.

12. Berkman and Plutzer, *Evolution, Creationism, and the Battle to Control America's Classrooms*, 78. Emphasis in original.

13. See, e.g., Adam Laats, *The Other School Reformers: Conservative Activism in American Education* (Cambridge, MA: Harvard University Press, 2015), 30–31.

14. "Special Report: Darwinian Theory Stirs Up Kentucky," *New York Times*, February 2, 1922.

15. "Editorial: Stop and Think," *Louisville Courier-Journal*, February 3, 1922.

16. *Moody Bible Institute Monthly* 21 (February 1921): 268.

17. William Jennings Bryan, "God and Evolution," *New York Times*, February 26, 1922; William Jennings Bryan, *The Bible and Its Enemies: An Address Delivered at the Moody Bible Institute of Chicago* (Chicago: Bible Institute Colportage Association, 1921), 23.

CHAPTER ONE

1. Birge to Conklin, March 4, 1922; Edwin A. Birge Papers, State Historical Society of Wisconsin Archives, Madison.

2. Irvin G. Wyllie, "Bryan, Birge, and the Wisconsin Evolution Controversy, 1921–1922," *Wisconsin Magazine of History* 35 (1951–52): 295.

3. Edward A. Birge, "Letter to the Editor," *Capital Times* (Madison, WI), February 17, 1922. Clipping in Birge Papers.

4. Ibid.

5. *Milwaukee Journal*, June 9, 1921. Clipping in Birge Papers.

6. *Milwaukee Sentinel*, June 9, 1921. Clipping in Birge Papers.

7. Wyllie, "Bryan, Birge, and the Wisconsin Evolution Controversy," 297.

8. See Michael Kazin, *A Godly Hero: The Life of William Jennings Bryan* (New York: Anchor Books, 2007), and Lawrence W. Levine, *Defender of the Faith: William Jennings Bryan; The Last Decade, 1915–1925* (New York: Oxford University Press, 1965).

9. William Jennings Bryan, "The Modern Arena," *Commoner* 21 (June 1921): 3.

10. Wyllie, "Bryan, Birge, and the Wisconsin Evolution Controversy," 298.

11. *Milwaukee Journal*, April 20, 1922. Clipping in Birge Papers.

12. *Milwaukee Journal*, April 17, 1922. Clipping in Birge Papers.

13. See M. T. Martin, "Letter to the Editor," *Capital Times* (Madison, WI), February 27, 1922; "Seedy," "Letter to the Editor," *Capital Times*, n.d.; and Lewis Ostenson, "Letter to the Editor," *Milwaukee Sentinel*, May 6, 1922. Clippings in Birge Papers.

14. John. W. Burgess, *Reminiscences of an American Scholar: The Beginnings of Columbia University* (New York: Columbia University Press, 1934), 147–48.

15. Noah Porter, quoted in Laurence R. Veysey, *The Emergence of the American University* (Chicago: University of Chicago Press, 1965), 44.

16. Ibid., 2.

17. Roger L. Geiger, *To Advance Knowledge: The Growth of American Research Universities, 1900–1940* (New York: Oxford University Press, 1986), 9.

18. John M. Coulter, *The Work of a University: Inaugural Address at Lake Forest, Ill.* (Madison, WI: Tracy, Gibbs, 1894), 4.

19. Geiger, *To Advance Knowledge*, 235.

20. Veysey, *Emergence of the American University*, 2.

21. John R. Thelin, *A History of American Higher Education* (Baltimore: Johns Hopkins University Press, 2004), 90.

22. William Bell Riley, "Higher Criticism in College Libraries," *School and Church* 2 (January–March 1920): 298.

23. T. T. [Thomas Theodore] Martin, *The Evolution Issue* (Los Angeles: N.d., n.p.[from an address October 28, 1923]).

24. J. Frank Norris, "The Inspiration of the Scriptures: Sermon by the Pastor in Answer to a Book of Higher Criticism by Dr. John A. Rice of Southern Methodist University," *Searchlight* (May 21, 1921): 1.

25. Louis Finkelstein, ed., *Thirteen Americans: Their Spiritual Autobiographies* (New York: Institute for Religious and Social Studies, 1953), 52, 57–58.

26. Edwin Grant Conklin, *The Direction of Human Evolution* (New York: Charles Scribner's Sons, 1921), v–vi. Emphases in original.

27. Ibid., vii.

28. Edwin G. Conklin, "Problems of Evolution and Present Methods of Attacking Them," *American Naturalist* 46 (March 1912): 121–28.

29. Conklin, *Direction of Human Evolution*, 192.

30. Edwin G. Conklin, "The Factors of Organic Evolution from the Standpoint of Embryology," in *Footnotes to Evolution: A Series of Popular Addresses on the Evolution of Life*, ed. David Starr Jordan (New York: D. Appleton, 1919), 100–117.

31. Edwin Grant Conklin, "Bryan and Evolution," *New York Times*, March 5, 1922.

32. William Jennings Bryan, *The Bible and Its Enemies: An Address Delivered at the Moody Bible Institute of Chicago* (Chicago: Bible Institute Colportage Association, 1921), 39.

33. William Jennings Bryan, *In His Image* (n.p., n.d. [1922]), 65–66.

34. James Gilbert, *Redeeming Culture: American Religion in an Age of Science* (Chicago: University of Chicago Press, 1997), 25–28; Burton Livingston to Bryan, September 29, 1924; William Jennings Bryan Papers, Library of Congress, Washington, DC.

35. Bryan, *In His Image*, 69.

36. William Jennings Bryan, "God and Evolution," *New York Times*, February 26, 1922.

37. Mark Noll, *The Scandal of the Evangelical Mind* (Grand Rapids, MI: Eerdmans, 1994), 197; Jon H. Roberts, *Darwinism and the Divine in America: Protestant Intellectuals and Organic Evolution, 1859–1900* (Madison: University of Wisconsin Press, 1988), 41–42.

38. William Jennings Bryan, *The Menace of Darwinism* (Louisville: Pentecostal Publishing, n.d.), 10; Bryan, "God and Evolution."

39. Bryan, *In His Image*, 65–66.

40. Bryan, "God and Evolution."

41. Ronald L. Numbers, *Darwinism Comes to America* (Cambridge, MA: Harvard University Press, 1999), 24.

42. Ibid., 33.

43. Edward J. Larson, *Evolution: The Remarkable History of a Scientific Theory* (New York: Modern Library, 2006), 128.

44. Roberts, *Darwinism and the Divine in America*, 87.

45. Charles Hodge, *What Is Darwinism?* (New York: Scribner, Armstrong, 1874), 177.

46. Henry Ward Beecher, *Evolution and Religion* (Boston: Pilgrim Press, 1885), i.

47. Ibid., iii.

48. James McCosh, *The Religious Aspect of Evolution* (New York: Charles Scribner's, 1890), vii.

49. Ibid., viii.

50. Ibid., 71.

CHAPTER TWO

1. Edward J. Larson, *Summer for the Gods: The Scopes Trial and America's Continuing Debate over Science and Religion* (Cambridge, MA: Harvard University Press, 2001); see also Ronald L. Numbers, *Darwinism Comes to America* (Cambridge, MA: Harvard University Press, 1998), 76–91; Jeffrey P. Moran, *The Scopes Trial: A Brief History with Documents* (Boston: Bedford/St. Martin's Press, 2002); and Moran, "Reading Race into the Scopes Trial: African American Elites, Science, and Fundamentalism," *Journal of American History* 90 (December 2003): 891–911. Not only was the actual science of evolution forced out of the 1925 trial, but it has been pushed out of subsequent trials as well. See, e.g., Harvey Siegel, "Creationism, Evolution, and Education: The California Fiasco," *Phi Delta Kappan* 63, no. 2 (1981): 95–101.

2. Sue Hicks to Ira Hicks, quoted in Larson, *Summer for the Gods*, 130.

3. Bryan to Sue Hicks, quoted in James Gilbert, *Redeeming Culture: American Religion in an Age of Science* (Chicago: University of Chicago Press, 1997), 32.

4. Bryan to Howard A. Kelly, June 10, 1925; Bryan Papers, Library of Congress, Washington, DC.

5. J. Frank Norris to Bryan, June or July 1925; Bryan Papers.

6. George McCready Price, "Modern Scientific Discoveries," *Christian Fundamentals in School and Church* 5 (October–December 1922).

7. George McCready Price, "Modern Problems in Science and Religion," *Moody Bible Institute Monthly* 21 (February 1921): 256.

8. G. M. Price to Bryan, February 27, 1922; Bryan Papers.

9. Moran, *Scopes Trial*, 49.

10. Larson, *Summer for the Gods*, 132–33. And, as Adam R. Shapiro argues in *Trying Biology: The Scopes Trial, Textbooks, and the Antievolution Movement in American Schools* (Chicago: University of Chicago Press, 2013), 107–10, both sides made erroneous assumptions about Scopes's presumed guilt. If Scopes were accused of teaching *human* evolution based solely on his use of the 1914 textbook *Civic Biology*, Shapiro concludes, then Scopes should have walked free. The book's ambivalent content on the issue of humanity's origins did not violate Tennessee's 1925 law.

11. Larson, *Summer for the Gods*, 80–83, 87–92.

12. See Adam Laats, *The Other School Reformers: The Conservative Tradition in American Education* (Cambridge, MA: Harvard University Press, 2015), chap. 2.

13. Arthur Garfield Hays, quoted in Larson, *Summer for the Gods*, 189.

14. Larson, *Summer for the Gods*, 191.

15. Frank L. McVey, "Address to the People of Kentucky," *Journal of the Kentucky Senate 1922*, 1031.

16. "Brands Darwin Bill Foes Apes," *Louisville Courier-Journal*, February 15, 1922.

17. George W. Ellis to Bryan and J. W. Porter, March 13, 1922; Bryan Papers.

18. Kentucky House Bill 191, *Journal of the House of Representatives of the Commonwealth of Kentucky 1922*, 1668–69.

19. Amendment to Kentucky Senate Bill 136, *Journal of the Kentucky Senate 1922*, 1062.

20. Richard David Wilhelm, "A Chronology and Analysis of Regulatory Actions Relating to the Teaching of Evolution in Public Schools" (PhD diss., University of Texas–Austin, 1978), 323 [South Carolina 1922 Amendment].

21. Georgia House Resolution 58.246-B, *Journal of the House of Representatives of the State of Georgia 1923*, 353–54; Georgia House Resolution 93.390-C, *Journal of the House of Representatives of the State of Georgia 1923*, 553; Georgia House Resolution 93, *Journal of the House of Representatives of the State of Georgia 1923*, 1001–2; Wilhelm, "Chronology and Analysis," 324–25 [Texas House Bill 97]; Texas House Concurrent Resolution 6, *Journal of the House of Representatives of the Third Called Session of the Thirty-Eighth Legislature of Texas 1923*, 73–74, 83.

22. West Virginia House Bill 153, *Journal of the House of Delegates of the State of West Virginia 1923*, 743, 901, 1947; Tennessee House Bill 947, *Tennessee House Journal 1923*, 666, 694, 719; Alabama Senate Joint Resolution 55, *Alabama Senate Journal 1923*, 1211–12; Iowa House File 657, *State of Iowa 1923 Journal of the House*, 758, 1048.

23. Florida House Concurrent Resolution 7, *Journal of the House of Representatives of the State of Florida 1923*, 482–83, 1176, 1853–54, 1878, 2025–27, 2200–2201, 2320, 3187–90.

24. Oklahoma House Bill 197, *Oklahoma House Journal 1923*, 304–5; Oklahoma House Bill 197, *Oklahoma Senate Journal 1923*, 1718–20; Oklahoma House Bill 197, *Session Laws of Oklahoma 1923*, 296; G. W. Moothart to Bryan, December 5, 1922, Bryan Papers; J. J. Walters [Governor of Oklahoma] to Bryan, May 26, 1923, Bryan Papers; Kenneth K. Bailey, "The Antievolution Crusade of the Nineteen-Twenties" (PhD diss., Vanderbilt University, 1954), 74.

25. "Science and Religion," *New York Times*, April 5, 1925; Bailey, "Antievolution Crusade," 67–68.

26. Bailey, "Antievolution Crusade," 68–69.

27. Quotation from Congress's Washington, DC, teacher law is from "Evolution Battle to Go to Congress: New Law Is Sought," *New York Times*, July 24, 1925; *Congressional Record*, 68th Cong., 1st sess., vol. 65, pt. 8:7796.

28. West Virginia House Bill 175, *West Virginia Bills of the House of Delegates 1925*, 66; Texas House Bill 378, *Texas House Journal 1925*, 386; Florida House Bill 691, *Journal of the Florida House of Representatives 1925*, 1267, 1579–80; "Anti-Darwin Campaigns Stir South and West," *New York Times*, June 10, 1923.

29. Bailey, "Antievolution Crusade," 71–72.

30. Tennessee House Bill 185, *House Journal of the Sixty-Fourth General Assembly of the State of Tennessee 1925*, 180, 201, 210, 248, 261, 268, 648, 655, 741; Tennessee House Bill 185, *Senate Journal of the Sixty-Fourth General Assembly of the State of Tennessee 1925*, 516–17; "Tennessee Bans the Teaching of Evolution," *New York Times*, March 24, 1925; "Fights Evolution to Uphold Bible," *New York Times*, July 5, 1925.

31. Edwin Grant Conklin, "Bryan and Evolution," *New York Times*, March 5, 1922.

32. "Professors to War on Evolution Laws," *New York Times*, January 1, 1927.

33. Adam Laats, *Fundamentalism and Education in the Scopes Era: God, Darwin, and the Roots of America's Culture Wars* (New York: Palgrave Macmillan, 2010), 99–120.

34. *Congressional Record*, 69th Cong., 1st sess., vol. 67, pt. 5:5748 [Blanton]; "Evolution Battle to Go to Congress."

35. *Congressional Record*, 69th Cong., 1st sess.,vol. 67, pt. 5:5747–48 [Lehlbach]; 5749 [LaGuardia].

36. James C. Young, "College Professors, Scientists and Fundamentalists Prepare to Do Battle," *New York Times*, January 30, 1927; Kentucky House Bill 96, *Journal of the House of Representatives of the Commonwealth of Kentucky 1926*, 173; Louisiana House Bills 41, 208, 279, and 314, *Official Journal of the Proceedings of the House of Representatives of the State of Louisiana at the Third Regular Session of the Legislature 1926*, 11, 34, 99–100, 130, 167, 184, 190, 219, 574–75, 642–43, 675, 700, 762, 782, 833; Maynard Shipley, *The War on Modern Science* (New York: Alfred A. Knopf, 1927), 79.

37. Bailey, "Antievolution Crusade," 224–30, 247; Maynard Shipley, "A Year of the Monkey War," *Independent*, October 1, 1927, 326–28, 344–45; Young, "College Professors, Scientists and Fundamentalists Prepare to Do Battle"; "Protest in Florida on Anti-Darwin Bill," *New York Times*, April 30, 1927, 3; "Missouri Kills Anti-Evolution Bill," *New York Times*, February 9, 1927, 8; "Act on Evolution Bills," *New York Times*, February 10, 1927, 38; "Beat Anti-Evolution Bill," *New York Times*, April 14, 1927, 24.

38. Arkansas House Bill 34, *Journal of the House of Representatives for the General Assembly of the State of Arkansas 1927*, 68–69, 263, 323–24; Bailey, "Antievolution Crusade," 243–47; Virginia Gray, "Anti-Evolution Sentiment and Behavior: The Case of Arkansas," *Journal of American History* 62 (1970): 353–65; Shipley, "Year of the Monkey War," 326.

39. Alabama House Bill 30, *Journal of the House of Representatives of the State of Alabama 1927*, 89; Wilhelm, "Chronology and Analysis," 368 [Alabama House Bill 30], 374 [South Carolina House Bill 60]; Arkansas House Bill 34, *Journal of the House of Representatives for the General Assembly of the State of Arkansas 1927*, 68–69, 263, 323–24.

40. North Dakota House Bill 222, *State of North Dakota Journal of the House of the Twentieth Session of the Legislative Assembly* 1927, 519, 1022.

41. West Virginia House Bill 264, *Journal of the House of Delegates of West Virginia* 1927, 104.

42. Florida House Bill 87, *Florida House Journal* 1927, 3000–3001; *Journal of the State Senate of Florida of the Session of* 1927, 2411–12.

43. Moran, *Scopes Trial*, 156.

44. William Jennings Bryan, "Who Shall Control?"; typescript statement in Bryan Papers, file 3, Dayton Trial Correspondence, June 24–30, 1925.

45. *Pierce v. Society of Sisters*, 268 U.S. 510 (1925).

46. Bryan, "Who Shall Control?"

47. Mary Balch Women's Christian Temperance Union to Bryan, May 26, 1925; Bryan Papers.

48. R. J. Alderman, "Evolution Leads to Sodom," *Moody Bible Institute Monthly* 23 (September 1922): 12.

49. Louis Finkelstein, ed., *Thirteen Americans: Their Spiritual Autobiographies* (New York: Institute for Religious and Social Studies, 1953), 55.

50. Ibid., 60.

51. Shipley, *War on Modern Science*, 3.

52. Ibid., 4.

53. Ibid., 3.

54. Russell Owen, quoted in Larson, *Summer for the Gods*, 201.

55. Maynard Shipley, "Evolution Still a Live Issue in the Schools," *Current History* 27 (March 1928): 801.

56. T. C. Horton, "Bryan the Brave—'Defender of the Faith,'" *King's Business* 16 (September 1925).

57. C. H. Thurber to Bryan, November 21, 1923; Bryan to Thurber, December 22, 1923; Bryan Papers.

CHAPTER THREE

1. Edward J. Larson, *Evolution: The Remarkable History of a Scientific Theory* (New York: Modern Library, 2006), 121–22.

2. Ibid., 160.

3. Philip Kitcher, *Living with Darwin: Evolution, Design, and the Future of Faith* (New York: Oxford University Press, 2007), 22, 75; Larson, *Evolution*, 226–227; Michael Ruse, *The Evolution-Creation Struggle* (Cambridge, MA: Harvard University Press, 2005), 172–77.

4. Larson, *Evolution*, 234.

5. Ruse, *Evolution-Creation Struggle*, 182–84 [Ford], 186–87 [Mayr].

6. Larson, *Evolution*, 247.

7. George Gaylord Simpson, *This View of Life: The World of an Evolutionist* (New York: Harcourt, Brace and World, 1964), vii.

8. Ibid., vii.

9. Ibid., 25.

10. John T. Scopes and James Presley, *Center of the Storm: Memoirs of John T. Scopes* (New York: Holt, Rinehart and Winston, 1967), 269.

11. Ibid., 271.

12. George Gaylord Simpson, "One Hundred Years without Darwin Are Enough," *Teachers College Record* 62, no. 8 (May 1961): 623.

13. Larry Cuban, *How Teachers Taught: Constancy and Change in American Classrooms, 1890–1980* (New York: Longman, 1984), 12; Arthur Zilversmit, *Changing Schools: Progressive Education Theory and Practice, 1930–1960* (Chicago: University of Chicago Press, 1993), 20–21.

14. George M. Marsden, *Fundamentalism and American Culture: The Shaping of Twentieth-Century Evangelicalism, 1870–1925* (New York: Oxford University Press, 1980), 191–94; Joel A. Carpenter, *Revive Us Again: The Reawakening of American Fundamentalism* (New York: Oxford University Press, 1997), 3.

15. Adam R. Shapiro, *Trying Biology: The Scopes Trial, Textbooks, and the Antievolution Movement in American Schools* (Chicago: University of Chicago Press, 2013), 113–14.

16. Ibid., 155.

17. Adam Laats, *Fundamentalism and Education in the Scopes Era: God, Darwin, and the Roots of America's Culture Wars* (New York: Palgrave Macmillan, 2010), 131–35.

18. LaDonna Robinson Olson, *Legacy of Faith: The Story of Bryan College* (Hayesville, NC: Schoettle, 1995), 49–50.

19. Ibid., 4.

20. Bob Jones Sr., *Original Intentions of the Founder* (Greenville, SC: Bob Jones University Press, 1960). Pamphlet in Bob Jones University Archives, Greenville, SC.

21. Laats, *Fundamentalism and Education in the Scopes Era*, 43–60, 121–38.

22. Frank Newport, "In U.S., 42% Believe Creationist View of Human Origins," Gallup.com, June 2, 2014, http://www.gallup.com/poll/170822/believe-creationist-view-human-origins.aspx?utm_source=alert&utm_medium=email&utm_campaign=syndication&utm_content=morelink&utm_term=All%20Gallup%20Headlines%20-%20Americas%20-%20Government%20-%20Politics%20-%20Science%20-%20USA (accessed March 24, 2015); Adam Laats, "Bad News for Everyone," *I Love You but You're Going to Hell*, June 3, 2014, http://iloveyoubutyouregoingtohell.org/2014/06/03/bad-news-for-everyone/ (accessed March 24, 2015).

23. Ronald L. Numbers, ed., *Creation-Evolution Debates* (New York: Garland, 1995), 160–61.

24. Ronald L. Numbers, *The Creationists: From Scientific Creationism to Intelligent Design*, exp. ed. (Cambridge, MA: Harvard University Press, 2006), 163; Numbers, *Creation-Evolution Debates*, 186.

25. Numbers, *Creationists*, 137–42.

26. Ibid., 8.

27. Simpson, *This View of Life*, vii.

28. Arnold B. Grobman, *The Changing Classroom: The Role of the Biological Sciences Curriculum Study* (New York: Doubleday, 1969), 1, 22.

29. Kenneth K. Bailey, "The Antievolution Crusade of the Nineteen-Twenties" (PhD diss., Vanderbilt University, 1953), 249.

30. Judith V. Grabiner and Peter D. Miller, "Effects of the Scopes Trial: Was It a Victory for Evolutionists?" *Science* 185 (September 6, 1974): 833.

31. George William Hunter, *Civic Biology* (New York: American Book Company, 1914), 195–96; George William Hunter, *New Civic Biology* (New York: American Book Company, 1926), 250–51.

32. Hunter, *Civic Biology* (1914), 422 [index entry on Darwin and natural selection]; Hunter, *New Civic Biology* (1926), 435 [index entry on Darwin and heredity].

33. "Texas Schoolbooks Omit Evolution," *New York Times*, July 13, 1926.

34. Truman Moon, *Biology for Beginners* (New York: Holt, 1921), v.

35. Truman Moon, *Biology for Beginners*, rev. ed. (New York: Holt, 1926), v.

36. Grabiner and Miller, "Effects of the Scopes Trial," 834.

37. "Texas Schoolbooks Omit Evolution."

38. Gerald Skoog, "The Coverage of Human Evolution in High School Biology Textbooks in the 20th Century and in Current State Standards," *Science and Education* 14 (2005): 401.

39. Ibid., 402–3.

40. Ella T. Smith, *Exploring Biology* (New York: Harcourt, Brace, 1938), 572.

41. Ibid., 528.

42. Ella Thea Smith, *Exploring Biology*, 4th ed. (New York: Harcourt, Brace, 1954), 465.

43. Ibid., 480.

44. Ella Thea Smith, *Exploring Biology: The Science of Living Things*, 5th ed. (New York: Harcourt, Brace, 1959), 583.

45. Ibid., 584.

46. Oscar Riddle, F. L. Fitzpatrick, H. B. Glass, B. C. Gruenberg, D. F. Miller, and E. W. Sinnott, eds., *The Teaching of Biology in Secondary Schools of the United States: A Report of Results from a Questionnaire* (Washington, DC: Union of American Biological Sciences, 1942), 7.

47. Ibid., 71. Emphasis in original.

48. Ibid., 76.

49. Ibid., 73–74.

50. Estelle R. Laba and Eugene W. Gross, "Evolution Slighted in High-School Biology," *Clearing House* 24 (March 1950): 396.

51. Ibid., 398.

52. Ibid., 397.

53. Ibid.

54. Jonathan Zimmerman, *Whose America? Culture Wars in the Public Schools* (Cambridge, MA: Harvard University Press, 2002), 135–59.

55. "Pressure from the Pulpit," *Evolution: A Journal of Nature* 3 (June 1937): 9.

56. L.E.K., "Editorial: Some Very Fancy Pussyfooting," *Evolution: A Journal of Nature* 4 (January 1938): 2.

57. Simpson, *This View of Life*, 26.

58. Grobman, *Changing Classroom*, 1.

59. Edward J. Larson, *Trial and Error: The American Controversy over Creation and Evolution*, 3rd. ed. (New York: Oxford University Press, 2003), 95–96.

60. Skoog, "Coverage of Human Evolution," 404.

61. Dorothy Nelkin, *Science Textbook Controversies and the Politics of Equal Time* (Cambridge, MA: MIT Press, 1978), 30.

62. Skoog, "Coverage of Human Evolution," 405; Grobman, *Changing Classroom*, 211.

63. Laats, *Fundamentalism and Education in the Scopes Era*, 91–95.

64. *Engel v. Vitale*, 370 U.S. 421 (1962).

65. *Abington School Dist. v. Schempp*, 374 U.S. 203 (1963); Donald E. Boles, *The Bible, Religion, and the Public Schools* (Ames: Iowa State University Press, 1963), 53. The only states that banned Bible reading in public schools before 1963 were Arizona, California, Illinois, Louisiana, Nebraska, Nevada, Ohio, South Dakota, Washington, and Wisconsin.

66. *Schempp*, 374 U.S. 203.

67. "Compulsory Devotions Banned; Bible Retains Classroom Value," *Christianity Today* 7 (July 5, 1963): 26.

68. "Education and the Evangelical Minority," *Christianity Today* 8 (February 28, 1964): 24 [502]–26.

CHAPTER FOUR

1. Duane T. Gish, "The Scientific Case for Creation," in *Evolutionists Confront Creationists: Proceedings of the 63rd Annual Meeting of the Pacific Division, American Association for the Advancement of Science*, ed. Frank Awbrey and William Thwaites, vol. 1, pt. 3 (San Francisco: Pacific Division, American Association for the Advancement of Science, 1984), 26.

2. Duane T. Gish, *Teaching Creation Science in Public Schools* (El Cajon, CA: Institute for Creation Research, 1995), v. As Ronald Numbers has noted, Gish mistakenly attributed the line to Clarence Darrow. See Ronald L. Numbers, *Darwinism Comes to America* (Cambridge, MA: Harvard University Press, 1998), 91.

3. *Kitzmiller v. Dover* trial transcript, TalkOrigins Archive, http://www.talkorigins.org/faqs/dover/kitzmiller_v_dover_decision.html#p9 (accessed March 25, 2015).

4. Arnold B. Grobman, *The Changing Classroom: The Role of the Biological Sciences Curriculum Study* (New York: Doubleday, 1969), 213; Dorothy Nelkin, *Science Textbook Controversies and the Politics of Equal Time* (Cambridge, MA: MIT Press, 1978), 29.

5. Grobman, *Changing Classroom*, 205.

6. Nelkin, *Science Textbook Controversies*, 29.

7. Grobman, *Changing Classroom*, 210.

8. Judith V. Grabiner and Peter D. Miller, "Effects of the Scopes Trial: Was It a Victory for Evolutionists?," *Science* 185 (September 6, 1974): 836.

9. Grobman, *Changing Classroom*, 210.

10. Howard Seymour, quoted in ibid., 207.

11. Nelkin, *Science Textbook Controversies*, 81.

12. "Darwin's Foes Lose Coast School Fight," *New York Times*, January 10, 1964.

13. Ronald L. Numbers, *The Creationists: From Scientific Creationism to Intelligent Design*, exp. ed. (Cambridge, MA: Harvard University Press, 2006), 270–71.

14. Richard David Wilhelm, "A Chronology and Analysis of Regulatory Actions Relating to the Teaching of Evolution in Public Schools" (PhD diss., University of Texas–Austin, 1978), 64.

15. Ibid., 281.

16. Ibid., 436.

17. Ibid., 79.

18. Ibid., 79.

19. Edward J. Larson, *Trial and Error: The American Controversy over Creation and Evolution*, 3rd ed. (New York: Oxford University Press, 2003), 135.

20. Wilhelm, "Chronology and Analysis," 281.

21. Larson, *Trial and Error*, 143–44.

22. Larson, *Trial and Error*, 237.

23. Ibid., 128–30.

24. George Gaylord Simpson, *This View of Life: The World of an Evolutionist* (New York: Harcourt, Brace and World, 1964), 25. [From an article published in *Science* 121 (April 1, 1960): 966–97).] For another example of the popular "Santa Claus" analogy, see Larson, *Trial and Error*, 192.

25. Theodosius Dobzhansky, "Nothing in Biology Makes Sense except in the Light of Evolution," *American Biology Teacher* 35 (1973): 125–29.

26. "A Statement Affirming Evolution as a Principle of Science," *Humanist* (January–February 1977): 4–6.

27. Grabiner and Miller, "Effects of the Scopes Trial."

28. Wilhelm, "Chronology and Analysis," 156–57.

29. Larson, *Trial and Error*, 106, 110; Wilhelm, "Chronology and Analysis," 157.

30. Wilhelm, "Chronology and Analysis," 155.

31. Nelkin, *Science Textbook Controversies*, 51.

32. Larson, *Trial and Error*, 98–99.

33. Susan Epperson, quoted in Susanna McBee and John Neary, "Evolution Revolution in Arkansas," *Life*, November 22, 1968, 89.

34. Philip J. Hirschkop, "Brief of the National Education Association of the United States and the National Science Teachers Association as Amici Curiae," *Epperson v. Arkansas*, 393 US 97 (1968), 4.

35. Ibid., 13.

36. *Epperson v. Arkansas*, 393 US 97 (1968).

37. Numbers, *Creationists*, 352.

38. Wendell R. Bird, "Freedom of Religion and Science Instruction in Public Schools," *Yale Law Journal* 87 (January 1978): 517.

39. Ibid., 518.

40. Richard M. Bliss, *Origins: Two Models: Evolution, Creation* (San Diego, CA: Creation-Life, 1976), 31. Emphasis in original.

41. Henry M. Morris, *Christian Education for the Real World* (El Cajon, CA: Master Books, 1991), 228.

42. Larson, *Trial and Error*, 150–51, 153; Numbers, *Creationists*, 352.

43. "Act 590 of 1981: General Acts, 73rd General Assembly, State of Arkansas," in *Creationism, Science and the Law: The Arkansas Case*, ed. Marcel C. LaFollette (Cambridge, MA: MIT Press, 1983), 15.

44. "Opinion of William R. Overton," *McLean v. State of Arkansas*, 663 F.2d 47, 48 (8th Cir. Ct. App. Ark., 1981), in LaFollette, *Creationism, Science and the Law*, 38; Marcel C. LaFollette, "Creationism in the News: Mass Media Coverage of the Arkansas Trial," in LaFollette, *Creationism, Science and the Law*, 188–207.

45. Larson, *Trial and Error*, 166–84; *Edwards v. Aguillard*, 482 U.S. 578 (1987).

46. Larson, *Trial and Error*, 169; Stuart Taylor Jr., "High Court Voids Curb on Teaching Evolution Theory," *New York Times*, June 20, 1987.

47. Kenneth M. Pierce, "Education: Putting Darwinism Back in the Dock," *Time*, March 16, 1981, http://www.time.com/time/magazine/article/0,9171,922471,00.html (accessed March 25, 2015).

48. "Around the Nation," *New York Times*, November 18, 1981.

49. "Public Favorable to Creationism," February 14, 2001, Gallup.com, http://www.gallup.com/poll/2014/Public-Favorable-Creationism.aspx (accessed March 25, 2015).

50. Numbers, *Creationists*, 373.

51. Percival Davis and Dean H. Kenyon, *Of Pandas and People: The Central Question of Biological Origins*, 2nd ed. (Dallas: Haughton, 1993), 12.

52. Numbers, *Creationists*, 375–76.

53. Larson, *Trial and Error*, 195; Numbers, *Creationists*, 383.

54. "Department Position on Evolution and 'Intelligent Design,'" Lehigh University Department of Biological Sciences website, http://www.lehigh.edu/~inbios/news/evolution.htm (accessed March 25, 2015).

55. Numbers, *Creationists*, 381–82.

56. Ibid., 391–92; Neela Banerjee, "An Alternative to Evolution Splits a Pennsylvania Town," *New York Times*, January 16, 2005.

57. "Excerpt from the Ruling on Intelligent Design," *New York Times*, December 21, 2005; Laurie Goodstein, "Issuing Rebuke, Judge Rejects Teaching of Intelligent Design," *New York Times*, December 21, 2005.

58. Laurie Goodstein, "Evolution Slate Outpolls Rivals," *New York Times*, November 9, 2005.

59. Davis and Kenyon, *Of Pandas and People*, viii.

60. Michael J. Behe, "Whether Intelligent Design Is Science: A Response to the Opinion of the Court in Kitzmiller vs. Dover Area School District" (Seattle: Discovery Institute, 2006): http://www.discovery.org/scripts/viewDB/filesDB-download.php?command=download&id=697 (accessed March 25, 2015).

61. Guillermo Gonzalez, "More Propaganda about Intelligent Design," *Ames (IA) Tribune*, February 9, 2006, available via http://www.discovery.org/a/3234 (accessed March 25, 2015).

62. David Klinghoffer, "What's the Matter with Kansas? Dishonest Darwinists—Coming to a State Near You," *National Review*, August 3, 2006, available via http://www.discovery.org/a/3686 (accessed March 25, 2015).

63. Jodi Wilgoren, "Kansas Board Approves Challenges to Evolution," *New York Times*, November 9, 2005.

64. Glenn Branch and Eugenie Scott, "The Antievolution Law That Wasn't," *American Biology Teacher* 65 (2003): 165–66.

65. Michael B. Berkman and Eric Plutzer, "Defeating Creationism in the Courtroom, but Not in the Classroom," *Science* 331 (January 28, 2011): 404–5.

CHAPTER FIVE

1. What follows draws to varying degrees on Harvey Siegel, "Creationism, Evolution, and Education: The California Fiasco," *Phi Delta Kappan* 63, no. 2 (1981): 95–101; Siegel, "The Response to Creationism," *Educational Studies* 15, no. 4 (1984): 349–64; and Mike U. Smith, Harvey Siegel, and J. D. McInerney, "Foundational Issues in Evolution Education," *Science and Education* 4, no. 1 (1995): 23–46.

2. These criteria were widely thought to be sufficient to demarcate genuine science from pseudoscience; a representative discussion is that of Carl G. Hempel, *Philosophy of Natural Science* (Englewood Cliffs, NJ: Prentice-Hall, 1966). According to philosopher Karl R. Popper, only falsifiability counts: that is, a theory counts as scientific only if it is capable of being shown to be false. The classic statement of Popper's view is his *The Logic of Scientific Discovery* (New York: Routledge, 1959).

3. The qualification "in principle" is needed because hypotheses might be confirmable in principle but not practice owing to technological or other practical limitations. For example, "There are canal-like structures on Mars" has always been confirmable in principle, but became confirmable in practice only when adequate telescopes were developed. It is confirmability-in-principle that is required for scientific status on the standard view.

4. Or retrodict, i.e., make and confirm predictions concerning the past.

5. This is a whitewashed version of the actual history, used here just to illustrate the relevant ideas. For more historically accurate accounts of the example and the controversies it raised, cf. B. Almassi, "Trust in Expert Testimony: Eddington's 1919 Eclipse Expedition and the British Response to General Relativity," *Studies in History and Philosophy of Modern Physics* 40, no. 1 (2009): 57–67, and D. Kennefick, "Testing Relativity from the 1919 Eclipse—a Question of Bias," *Physics Today* 62, no. 3 (2009): 37–42. Thanks here to Otávio Bueno for helpful discussion.

6. The phrase was introduced by Donald T. Campbell in his article "Blind Variation and Selective Retentions in Creative Thought as in Other Knowledge Processes," *Psychological Review* 67, no. 6 (1960): 380–400.

7. In addition to the raft of biology textbooks available on the market, helpful introductions to the theory of evolution and the philosophical issues it raises include A. Rosenberg and D. W. McShea, *Philosophy of Biology: A Contemporary Introduction* (New York: Routledge, 2008); Elliot Sober, *Philosophy of Biology*, 2nd ed. (Boulder, CO: Westview Press, 2000); and Michael Ruse, ed., *Philosophy of Biology* (New York: Macmillan, 1989).

8. Thanks here to Peter Luykx for helpful conversation and advice.

9. It is also sometimes suggested that evolutionary theory is not scientific because it deals with the past—"origins"—and the past is beyond our observational capabilities and so not open to scientific investigation. For discussion and references, see Siegel, "Creationism, Evolution, and Education." There is of course a very large literature concerning the scientific status of both evolutionary theory and creationism. For a representative sample, cf. Niles Eldredge,

"Creationism Isn't Science," *New Republic* 4 (April 1981): 15–20; Philip Kitcher, *Abusing Science: The Case against Creationism* (Cambridge, MA: MIT Press, 1982); Robert T. Pennock, *Tower of Babel: The Evidence against the New Creationism* (Cambridge, MA: MIT Press, 1999); Robert T. Pennock, ed., *Intelligent Design Creationism and Its Critics: Philosophical, Theological and Scientific Perspectives* (Cambridge, MA: MIT Press, 2001); Robert T. Pennock and Michael Ruse, eds., *But Is It Science?: The Philosophical Question in the Creation/Evolution Controversy*, updated ed. (Amherst, NY: Prometheus Books, 2009); Ruse, *Philosophy of Biology*; Siegel, "Response to Creationism"; Sober, *Philosophy of Biology*; and Rosenberg and McShea, *Philosophy of Biology*.

10. Thanks here to the forceful comments of Michael Veber. As Randall Curren has pointed out, if the creationist posits *purpose* to the Creator, the same dilemma arises: for example, the existence of vestigial organs either counts as disconfirmation of purposeful creation or it does not, in which case creationism is unfalsifiable.

11. Darwin himself made this point. Cf. Charles Darwin, *On the Origin of Species by Means of Natural Selection, or the Preservation of Favoured Races in the Struggle for Life* (London: John Murray, 1859), 167, and Robert Nola, "Darwin's Arguments in Favour of Natural Selection and against Special Creation," *Science and Education* 22, no. 2 (2013): 149–71, for excellent discussion. As we will see in chapter 6, the point applies to ID as well.

12. This is not the only way in which the standard view has been superseded in recent decades, but it is the one most relevant to the present topic. Not all philosophers have conceded the insolubility of the demarcation problem; for a robust defense of demarcation, see Robert T. Pennock, "Can't Philosophers Tell the Difference between Science and Religion?: Demarcation Revisited," *Synthese* 178 (2011): 177–205. (First published in Pennock and Ruse, *But Is It Science?*) A plausible "family resemblance" view of the nature of science, with direct application to science education, is offered in Gurol Irzik and Robert Nola, "A Family Resemblance Approach to the Nature of Science for Science Education," *Science and Education* 20, nos. 7–8 (2011): 591–607. Several important early papers on the demarcation issue are reprinted in Pennock and Ruse, *But Is It Science?*

13. As demonstrated in Philip Kitcher, *Living with Darwin: Evolution, Design, and the Future of Faith* (New York: Oxford University Press, 2007), this is exactly what has happened to ID.

14. Cf. Kitcher, *Abusing Science*; Philip Kitcher, "Good Science, Bad Science, Dreadful Science, and Pseudo-Science," *Journal of College Science Teaching* 14 (1985): 168–73; and Kitcher, *Living with Darwin*; other worthwhile sources include several of the essays, especially those by Elliott Sober and Robert Pennock, in Pennock and Ruse, *But Is It Science?* It is worth noting that the vagueness of the boundary between science and non-science does not vitiate the distinction; presumably some things—culinary recipes, comic books, acts of kindness, etc.—can be ruled out as non-science, despite the just-mentioned vagueness. This point need not be pursued further here.

15. It is perhaps worth noting that secular humanists needn't be atheists; they can just as well be agnostics, and even theists, albeit of a nontraditional sort.

16. Creationists might nevertheless object to what they take to be the religious implications of evolutionary theory; in particular, they might (and do) object to the teaching of evolution on

the grounds that belief in it undermines their children's belief in incompatible religious theses. We address this in chapter 7.

17. For more on neutrality, see Naomi Oreskes and Erik M. Conway, *Merchants of Doubt: How a Handful of Scientists Obscured the Truth on Issues from Tobacco Smoke to Global Warming* (New York: Bloomsbury, 2010). Thanks to Mike U. Smith for drawing this book to our attention.

CHAPTER SIX

1. The relation between creationism and ID is in some ways obscure. Some writers think of ID as a strategic variant of creationism, a theologically cleansed Trojan Horse designed to get creationism into the curriculum on the basis of its nonreligious/scientific character. Others regard it as a distinct anti-evolution theory. We do not here address the strategies or motives of those advancing ID as a suitable classroom alternative to evolution. Instead, we treat it on its own terms and assess its suitability for the science classroom.

2. Philip Kitcher, *Living with Darwin: Evolution, Design, and the Future of Faith* (New York: Oxford University Press, 2007), p. 7.

3. William Paley, *Natural Theology; or, Evidences of the Existence and Attributes of the Deity* (Oxford: Oxford University Press, 2006; first published 1802).

4. Ibid., 8.

5. David Hume, *Dialogues concerning Natural Religion*, ed. R. H. Popkin (Indianapolis: Hackett, 1980; first published 1779).

6. Paley, *Natural Theology*, 35.

7. Ibid., 229. Emphasis in original.

8. Cf. Kitcher, *Living with Darwin*, 36–38, 59–60, and 100–113, as well as several of the essays in Robert T. Pennock, ed., *Intelligent Design Creationism and Its Critics: Philosophical, Theological and Scientific Perspectives* (Cambridge, MA: MIT Press, 2001), and Robert T. Pennock and Michael Ruse, eds., *But Is It Science?: The Philosophical Question in the Creation/Evolution Controversy*, updated ed. (Amherst, NY: Prometheus Books, 2009).

9. Cf. the works cited in the previous note for a healthy range of examples. A particularly accessible explanation of the evolution of vision and the eye may be seen in episode 2 of Neil deGrasse Tyson's PBS television series *Cosmos*. See also William Ayliffe, "The Evolution of Vision," www.youtube.com/watch?v=-JMyWerdguQ (accessed April 4, 2015).

10. With respect to explanatory power, see, in addition to the arguments above, Kitcher, *Living with Darwin*, 46–48, 61–64, 172n31.

11. Ibid., 113–14; cf. 8–12. Cf. also Robert T. Pennock, *Tower of Babel: The Evidence against the New Creationism* (Cambridge, MA: MIT Press, 1999), and the papers in Pennock, *Intelligent Design Creationism and Its Critics*, and Pennock and Ruse, *But Is It Science?*

12. Michael J. Behe, *Darwin's Black Box* (New York: Free Press, 1996).

13. In addition to ibid., see also Michael J. Behe, *The Edge of Evolution: The Search for the Limits of Darwinism* (New York: Free Press, 2007); William Dembski, *The Design Inference* (Cambridge: Cambridge University Press, 1998); and several of the papers in Pennock, *Intelligent Design Creationism and Its Critics*, and Pennock and Ruse, *But Is It Science?*

14. Thomas Nagel, "Public Education and Intelligent Design," *Philosophy and Public Affairs* 36, no. 2 (2008): 187–205.

15. Nagel makes his case more fully in his *Mind and Cosmos: Why the Materialist Neo-Darwinian Conception of Nature Is Almost Certainly False* (New York: Oxford University Press, 2012). This book has generated intense controversy and criticism; for a sample of critical reactions see John Dupré, "Review of Thomas Nagel, *Mind and Cosmos: Why the Materialist Neo-Darwinian Conception of Nature Is Almost Certainly False*," *Notre Dame Philosophical Reviews*, 2012, http://ndpr.nd.edu/news/35163/ (accessed April 4, 2015); Philip Kitcher, "Things Fall Apart" (Review of Thomas Nagel, *Mind and Cosmos: Why the Materialist Neo-Darwinian Conception of Nature Is Almost Certainly False*), *New York Times Opinionator*, http://opinionator.blogs.nytimes.com/2013/09/08/things-fall-apart/?_r=0 (accessed April 13, 2015); Colin McGinn, "Review of Thomas Nagel, *Mind and Cosmos: Why the Materialist Neo-Darwinian Conception of Nature Is Almost Certainly False*," *Mind* 122, no. 486 (2013): 582–85; H. Allen Orr, "Awaiting a New Darwin" (Review of Thomas Nagel, *Mind and Cosmos: Why the Materialist Neo-Darwinian Conception of Nature Is Almost Certainly False*), *New York Review of Books* 60, no. 2 (2013): 26–28; and Elliot Sober, "Remarkable Facts: Ending Science As We Know It" (Review of Thomas Nagel, *Mind and Cosmos: Why the Materialist Neo-Darwinian Conception of Nature Is Almost Certainly False*), *Boston Review*, 2012, http://bostonreview.net/books-ideas/remarkable-facts (accessed April 4, 2015).

16. Nagel, "Public Education and Intelligent Design," 187.

17. Ibid., 188.

18. Ibid.

19. Ibid., 202.

20. Ibid., 190–91.

21. Ibid., 192.

22. Behe, *Edge of Evolution*, 165.

23. Nagel, "Public Education and Intelligent Design," 192.

24. Ibid., 192.

25. Behe, *Edge of Evolution*, 167.

26. Nagel, "Public Education and Intelligent Design," 195, 199.

27. Edward Erwin and Harvey Siegel, "Is Confirmation Differential?," *British Journal for Philosophy of Science* 40, no. 1(1989): 105–19.

28. A confidence, we should note, that Nagel finds unwarranted. Cf. Nagel, "Public Education and Intelligent Design," 190–91, 199.

29. Kitcher, *Living with Darwin*, 82.

30. Behe, *Darwin's Black Box*, 94.

31. There is now a wealth of information and a refined understanding of the evolution of vision and the eye. See, e.g., "Evolution of the Eye," http://www.pbs.org/wgbh/evolution/library/01/1/l_011_01.html (accessed April 6, 2015); Trevor D. Lamb, "Evolution of the Eye," *Scientific American* 305, no. 1 (July 2011): 64–69.

32. For more detail, cf. Kitcher, *Living with Darwin*, 80–90, on which our discussion is based. For an account of the evolution of the flagellum, embedded in a broad account of the controversy, see "Intelligent Design on Trial," available at http://www.pbs.org/wgbh/nova/evolution/intelligent-design-trial.html (accessed April 4, 2015). For a remarkably simple and

effective demonstration of the increasing functionality of allegedly "irreducibly complex" things, see Kenneth Miller's demonstration of the functionality of mousetrap parts: https://www.youtube.com/watch?v=rW_2lLG9EZM (accessed April 4, 2015).

33. For example, Behe, *Darwin's Black Box*, 94.

34. Here, too, we are indebted to Kitcher, *Living with Darwin*, 90–100.

CHAPTER SEVEN

1. We don't want to bore the reader by constantly repeating that by *belief* we mean "belief true," so we occasionally use the locution "belief in" in order to render the prose more readable. In all cases, we mean "belief true." Thanks here to Mike U. Smith. We also note here that the notion of *truth* must be handled with care. When applied to scientific theories, like evolutionary theory, we interpret it as including such qualifiers as "approximately true" and excluding such features as certainty. There is of course much to say about truth that we can't say here. For related discussion, see Harvey Siegel, "Knowledge, Truth and Education," in *Education, Knowledge, and Truth: Beyond the Postmodern Impasse*, ed. David Carr (London: Routledge, 1998), 19–36.

2. Some might also worry that independently of aims or intentions, the *effect* of evolution education is to produce belief. We don't pursue this worry here, both because the evidence for such an effect is modest and because it is the aim with which we are concerned. For discussion, see Brian Alters, "Apprehension and Pedagogy in Evolution Education," *Evolution: Education and Outreach* 3 (2010): 231–35. Thanks to Glenn Branch for raising this concern.

3. The following discussion is taken from and builds on the treatment of this topic developed in Mike U. Smith and Harvey Siegel, "Knowing, Believing, and Understanding: What Goals for Science Education?" *Science and Education* 13, no. 6 (2004): 553–82, and Smith and Siegel, "On the Relationship between Belief and Acceptance as Goals of Evolution Education: Ten Years Later," *Cultural Studies of Science Education*, forthcoming. We are grateful to Mike U. Smith for his insightful contributions to and guidance concerning both the general topic of evolution education and the specific question of the status of belief as an aim of such education.

4. Andrea A. diSessa, "Unlearning Aristotelian Physics: A Study of Knowledge-Based Learning," *Cognitive Science* 2 (1982): 37–75; quotation is from p. 38. Additional evidence that many very good college science students are "closet Aristotelians" is offered in C. A. Lambert, "Twilight of the Lecture," *Harvard Magazine* 114, no. 4 (2012): 23–27, which discusses the problems with lecture-style physics education made much of by Harvard physicist Eric Mazur.

5. DiSessa's "Unlearning Aristotelian Physics" launched an important research path that continues today. For a glimpse of the current state of play, see A. A. diSessa, "A Bird's Eye View of the 'Pieces' vs. 'Coherence' Controversy (From the 'Pieces' Side of the Fence)," In *International Handbook of Research on Conceptual Change*, ed. S. Vosniadou (New York: Routledge, 2008), 35–60.

6. As we are using these terms, *evidence* is a type or species of *reasons*, which is a more general category. Evidence—in science, generally empirical evidence—is one type of reason. But other sorts of reasons, including consistency/inconsistency with other theories, modal relations (e.g., considerations concerning possibility/necessity), the epistemic force of particular items of evidence, etc., are not themselves empirical but nevertheless are key to the epistemic evaluation

of candidate hypotheses and theories. So reasons, the broad category, can be characterized as *considerations that count in favor of or against* the items for which they are reasons; (empirical) evidence is one type of consideration. For further discussion of these notions and their relationship, see Harvey Siegel, "The Rationality of Science, Critical Thinking, and Science Education," *Synthese* 80, no. 1 (1989): 9–41; Siegel, "Truth, Thinking, Testimony and Trust: Alvin Goldman on Epistemology and Education," *Philosophy and Phenomenological Research* 72, no. 2 (2005): 345–66; and Siegel, "The Role of Reasons in Moral Education," in *Moral Education and Development: A Lifetime Commitment*, ed. D. J. deRuyter and S. Meidema (Rotterdam: Sense, 2011), 59–69.

We think that exposing students in a substantial way to the history and especially the philosophy of science is a particularly good way to help them understand the nature of reasons and evidence in science, since many of the issues at the center of philosophy of science—the nature of scientific evidence; the relation between evidence and theory; the evaluation of the strength of particular bits of evidence and the degree to which they support (or fail to support) particular hypotheses and theories; the role of experimentation in both generating and evaluating evidential strength; the "logic" of science and the strengths and weaknesses of inductivist/falsificationist/abductivist accounts of that logic; the importance of and the difficulties involved in drawing a science/pseudoscience distinction; and accounts of the general features of scientific knowledge and inquiry (absolutism, relativism, pluralism, dogmatism, fallibilism, tentativeness, etc.)—involve issues concerning the nature and role of reasons in science. An important and influential case for the centrality of history and philosophy of science for science education is Michael R. Matthews, *Science Teaching: The Role of History and Philosophy of Science*, rev. and exp. 20th anniversary ed. (New York: Routledge, 2014).

7. For further discussion of this point concerning belief's being a necessary condition of knowledge, see Siegel, "Knowledge, Truth and Education," and Smith and Siegel, "Knowing, Believing, and Understanding."

8. We do not defend this "internalist" view of justification, or that justification is a necessary condition of knowledge, here. See Siegel, "Knowledge, Truth and Education"; Siegel, "Truth, Thinking, Testimony and Trust"; and Emily Robertson, "The Epistemic Aims of Education," in *The Oxford Handbook of Philosophy of Education*, ed. H. Siegel (New York: Oxford University Press, 2009), 11–34.

9. There is of course much more to say concerning the nature of understanding than we can say here. For more, see Smith and Siegel, "Knowing, Believing, and Understanding"; Robertson, "Epistemic Aims of Education," 19–20; and esp. Catherine Z. Elgin, "From Knowledge to Understanding," in *Epistemology Futures*, ed. S. Hetherington (Oxford: Oxford University Press, 2006), 199–215, and Elgin, "Education and the Advancement of Understanding," reprinted in *Philosophy of Education: An Anthology*, ed. Randall Curren (Oxford: Blackwell, 2007), 417–22.

10. Cf. Jonathan E. Adler, *Belief's Own Ethics* (Cambridge, MA: MIT Press, 2002), and Simon J. Evnine, "Learning from One's Mistakes: Epistemic Modesty and the Nature of Belief," *Pacific Philosophical Quarterly* 82 (2001): 157–77.

11. Cf. Israel Scheffler, *Reason and Teaching* (Indianapolis: Hackett, 1989); Harvey Siegel, *Educating Reason: Rationality, Critical Thinking, and Education* (London: Routledge, 1988);

and Siegel, *Rationality Redeemed?: Further Dialogues on an Educational Ideal* (New York: Routledge, 1997).

12. For extended discussion of how a teacher might appropriately and productively address student beliefs in the science classroom in addition to Smith and Siegel, "Knowing, Believing, and Understanding," cf. Smith and Siegel, "On the Relationship between Belief and Acceptance"; Mike U. Smith, "Current Status of Research in Teaching and Learning Evolution: I. Philosophical/Epistemological Issues," *Science and Education* 19, nos. 6–8 (2010): 523–38; Mike U. Smith, "Current Status of Research in Teaching and Learning Evolution: II. Pedagogical Issues," *Science and Education* 19, nos. 6–8 (2010): 539–71; and Mike U. Smith, Harvey Siegel, and J. D. McInerney, "Foundational Issues in Evolution Education," *Science and Education* 4, no. 1 (1995): 23–46.

13. The *nature of science* (NOS) has received a great deal of attention in the science education literature of the last few decades. A serious discussion of it is unfortunately beyond the scope of this book. For a selective introduction, see Matthews, *Science Teaching*; Siegel, "Rationality of Science, Critical Thinking, and Science Education"; Harvey Siegel, "Incommensurability, Rationality and Relativism: In Science, Culture and Science Education," in *Incommensurability and Related Matters*, ed. Paul Hoyningen-Huene and Howard Sankey (Dordrecht, Netherlands: Springer, 2001), 207–24; Siegel, "The Bearing of Philosophy of Science on Science Education and *Vice Versa*: The Case of Constructivism," *Studies in History and Philosophy of Science* 35A.1 (2004): 185–98; Smith and Siegel, "Knowing, Believing, and Understanding"; and Smith and Siegel, "On the Relationship between Belief and Acceptance."

CHAPTER EIGHT

1. A good example is the recent advance in HIV/AIDS research/treatment, in which successful drug "cocktails" successfully take into account the evolution of the virus.

2. For discussion of some of the many different ways the term is used and the many different forms of it treated in the educational and philosophical literature, see Harvey Siegel, "Multiculturalism and the Possibility of Transcultural Educational and Philosophical Ideals," *Philosophy* 74 (1999): 387–409.

3. For example, James A. Banks, "Multicultural Education: Characteristics and Goals," in *Multicultural Education: Issues and Perspectives*, ed. J. A. Banks and C. A. M. Banks, 7th ed. (Hoboken, NJ: Wiley, 2010), 3–30; D. Bernal, "Using a Chicana Feminist Epistemology in Educational Research," *Harvard Educational Review* 68, no. 4 (1998): 555–83; Henry Giroux, "Insurgent Multiculturalism and the Promise of Pedagogy," in *Multiculturalism: A Critical Reader*, ed. David Theo Goldberg (Oxford: Blackwell, 1994), 325–43; G. Ladson-Billings, "Racialized Discourses and Ethnic Epistemology," in *Handbook of Qualitative Research*, ed. N. Denzin and Y. Lincoln, 2nd ed. (Thousand Oaks, CA: Sage, 2000), 257–77; and C. Mills, "Alternative Epistemologies," *Social Theory and Practice* 14, no. 3 (1988): 237–63.

4. David Theo Goldberg, "Introduction: Multicultural Conditions," in Goldberg, *Multiculturalism: A Critical Reader*, 1–41; quotation is from pp. 17–18.

5. Many others, especially those treating multiculturalism in social/political terms—including Eamonn Callan, *Creating Citizens: Political Education and Liberal Democracy* (Oxford: Oxford

University Press, 1997); Robert K. Fullinwider, "Multicultural Education: Concepts, Policies, and Controversies," in *Public Education in a Multicultural Society: Policy, Theory, Critique*, ed. Robert K. Fullinwider (Cambridge: Cambridge University Press, 1996), 3–22; Amy Gutmann, *Democratic Education*, rev. ed. (Princeton, NJ: Princeton University Press, 1999), and Gutmann, "Educating for Individual Freedom and Democratic Citizenship: In Unity and Diversity There Is Strength," in *The Oxford Handbook of Philosophy of Education*, ed. H. Siegel (New York: Oxford University Press, 2009), 409–27; Martha Nussbaum, *Cultivating Humanity: A Classical Defense of Reform in Liberal Education* (Cambridge, MA: Harvard University Press, 1998); and Charles Taylor, *Multiculturalism and "The Politics of Recognition,"* with commentary by Amy Gutmann (ed.), Steven C. Rockefeller, Michael Walzer, and Susan Wolf (Princeton, NJ: Princeton University Press, 1992)—have similarly suggested that multiculturalism depends on an underlying universalism. But they have not pursued in depth the epistemological issues thus raised.

6. Israel Scheffler, *Teachers of My Youth: An American Jewish Experience* (Dordrecht, Netherlands: Springer, 1994), 15; cf. 87–90.

7. We take the expression from Thomas Nagel, *The View from Nowhere* (Oxford: Oxford University Press, 1986).

8. This section draws on Harvey Siegel, *Rationality Redeemed?: Further Dialogues on an Educational Ideal* (New York: Routledge, 1997); Siegel, "Multiculturalism and the Possibility of Transcultural Educational and Philosophical Ideals"; Siegel, "Incommensurability, Rationality and Relativism: In Science, Culture and Science Education," in *Incommensurability and Related Matters*, ed. Paul Hoyningen-Huene and Howard Sankey (Dordrecht, Netherlands: Springer, 2001), 207–24; and Siegel, "Multiculturalism and Rationality," *Theory and Research in Education* 5, no. 2 (2007): 203–23.

9. We do not endorse this interpretation but offer it as a reconstructive hypothesis. For discussion, cf. Robert Pennock, "The Postmodern Sin of Intelligent Design Creationism," *Science and Education* 19 (2010): 757–78.

10. "WMS" is not an expression invented by us; it is an acronym used in much of the multicultural science education literature. Cf., for example, William B. Stanley and Nancy W. Brickhouse, "Teaching Sciences: The Multicultural Question Revisited," *Science Education* 85, no. 1 (2001): 35–49; quotation is from p. 35.

11. See, e.g., R. Barnhardt and A. O. Kawagley, "Indigenous Knowledge Systems and Education," in *Why Do We Educate?: Renewing the Conversation (107th Yearbook of the NSSE, vol. 1)*, ed. D. L. Coulter and J. R. Wiens (Malden, MA: Blackwell, 2008), 223–41; D. Hodson, "In Search of a Rationale for Multicultural Science Education," *Science Education* 77, no. 6 (1993): 685–711; A. O. Kawagley, D. Norris-Tull, and R. Norris-Tull, "The Indigenous Worldview of Yupiaq Culture: Its Scientific Nature and Relevance to the Practice and Teaching of Science," *Journal of Research in Science Teaching* 35, no. 2 (1998): 133–44; M. Ogawa, "Science Education in a Multicultural Perspective," *Science Education* 79 (1995): 583–93; William B. Stanley and Nancy W. Brickhouse, "Multiculturalism, Universalism, and Science Education," *Science Education* 78, no. 4 (1994): 387–98; Stanley and Brickhouse, "Teaching Sciences"; and G. Snively and J. Corsiglia, "Rediscovering Indigenous Science: Implications for Science Education," *Science Education* 85, no. 1 (2001): 6–34.

12. In addition to the references noted immediately above in the text, cf. references and discussion in Michael R. Matthews, *Science Teaching: The Role of History and Philosophy of Science*, rev. and exp. 20th anniversary ed. (New York: Routledge, 2014), and Gurol Irzik and Sibel Irzik, "Which Multiculturalism?," *Science and Education* 11 (2002): 393–403.

13. Snively and Corsiglia, "Rediscovering Indigenous Science"; Charbel Niño El-Hani and Fábio Pedro Souza de Ferreira Bandeira, "Valuing Indigenous Knowledge: To Call It 'Science' Will Not Help," *Cultural Studies of Science Education* 3 (2008): 751–79.

14. Cf. Harvey Siegel, "Science Education: Multicultural *and* Universal," *Interchange* 28, no. 2 (1997): 97–108, 101–4; Siegel, "Incommensurability, Rationality, and Relativism," 216–20; and Siegel, "Multiculturalism, Universalism, and Science Education: In Search of Common Ground," *Science Education* 86, no. 6 (2002): 803–20, 810–19.

15. Southerland's (S. A. Southerland, "Epistemic Universalism and the Shortcomings of Curricular Multicultural Science Education," *Science and Education* 9, no. 3 [2000]: 289–307) very helpful distinction between *curricular* and *instructional* MSE is a propos here: the former treats WMS and "ethnic sciences" as being on an epistemic par; the latter focuses on the ways we teach and treat students from nonmajority cultures. As should be clear, we reject the former but embrace the latter form of MSE.

16. This limited, narrow, "bounded" superiority is well articulated in ibid., 295–96, 300.

17. Mike U. Smith and Harvey Siegel, "Knowing, Believing, and Understanding: What Goals for Science Education?" *Science and Education* 13, no. 6 (2004); Mike U. Smith and Harvey Siegel, "On the Relationship between Belief and Acceptance of Evolution as Goals of Evolution Education: Ten Years Later," *Cultural Studies of Science Education* 2015 (forthcoming).

18. This is, needless to say, problematic. For a brief case against, see Harvey Siegel, "The Response to Creationism," *Educational Studies* 15, no. 4 (1984): 360–62. Randall Curren, "Education as a Social Right in a Diverse Society," *Journal of Philosophy of Education* 41, no. 1 (2009): 45–56; Curren, "Right to an Education," in *Encyclopedia of Educational Theory and Philosophy*, ed. D. C. Phillips (London: Sage, 2014), 2:712–14; and Rob Reich, "Educational Authority and the Interests of Children," in Siegel, *Oxford Handbook of Philosophy of Education*, 469–85, offer systematic discussions of children's rights to education and of the legal and moral limitations of parents' rights to control their children's education. Robert Audi, "Science Education, Religious Toleration, and Liberal Neutrality toward the Good," in Siegel, *Oxford Handbook of Philosophy of Education*, 333–57, offers a nuanced discussion of the place of religious toleration in determining the science curriculum.

19. As noted earlier, determining the worthiness or otherwise of proposed curriculum content is at least in part a matter of its *disciplinary adequacy*: the degree to which it satisfies the current legitimate standards of the relevant discipline.

20. For consideration of the ways in which the understanding of ET can contribute in practical ways to the lives of students, see Mike U. Smith, "Current Status of Research in Teaching and Learning Evolution: II. Pedagogical Issues," *Science and Education* 19, nos. 6–8 (2010): 539–71; and Smith and Siegel, "On the Relationship between Belief and Acceptance."

21. The expression is Joseph Schwab's; for discussion and references, cf. Harvey Siegel, *Educating Reason: Rationality, Critical Thinking, and Education* (London: Routledge, 1988), chap. 6.

22. For more general discussion of the aims of science education in general and biology education in particular, cf. ibid., chap. 6; Smith and Siegel, "Knowing, Believing, and Understanding"; and Smith and Siegel, "On the Relationship between Belief and Acceptance."

23. Mary Midgley, *Intelligent Design Theory and Other Ideological Problems*, Impact Series 15 (London: Philosophy of Education Society of Great Britain, 2007), 43.

24. Siegel, *Educating Reason*. Concerning the treatment of students from "anti-evolution" or "creationist" cultures, in addition to Southerland, "Epistemic Universalism and the Shortcomings of Curricular Multicultural Science Education," see also William W. Cobern and C. C. Loving, "Defining 'Science' in a Multicultural World: Implications for Science Education," *Science Education* 85 (2001): 50–67; Smith, "Current Status of Research in Teaching and Learning Evolution: II. Pedagogical Issues"; Charbel N. El-Hani and E. F. Mortimer, "Multicultural Education, Pragmatism, and the Goals of Science Teaching," *Cultural Studies of Science* 2 (2007): 657–702; and El-Hani and Bandeira, "Valuing Indigenous Knowledge." El-Hani and colleagues explain and defend "culturally-sensitive science education," in which science is taught "authoritatively" but sensitively. We think their general approach to such education can and should be applied to the evolution vs. creationism/ID controversy when that controversy is understood as a cultural rather than a scientific matter.

CONCLUSION

1. Philip Kitcher, *Living with Darwin: Evolution, Design, and the Future of Faith* (New York: Oxford University Press, 2007), 8.

2. Michael Berkman and Eric Plutzer, *Evolution, Creationism, and the Battle to Control America's Classrooms* (New York: Cambridge University Press, 2010).

3. Michael Lienesch, *In the Beginning: Fundamentalism, the Scopes Trial, and the Making of the Antievolution Movement* (Chapel Hill: University of North Carolina Press, 2007), 5.

4. See, e.g., Carl E. Kaestle, *Pillars of the Republic: Common Schools and American Society, 1780–1860* (New York: Hill and Wang, 1983), 162–71, and James C. Carper and Thomas C. Hunt, *The Dissenting Tradition in American Education* (New York: Peter Lang, 2007), 11–28.

5. Steven K. Green, *The Bible, the School, and the Constitution: The Clash That Shaped Modern Church-State Doctrine* (New York: Oxford University Press, 2012), 254–57.

6. See, e.g., Lisa Delpit, *Other People's Children: Cultural Conflict in the Classroom*, 2nd ed. (New York: Free Press, 2005).

7. See, for example, Mike U. Smith and Harvey Siegel's 2004 suggestion of the understanding-not-belief approach (Smith and Siegel, "Knowing, Believing, and Understanding: What Goals for Science Education?," *Science and Education* 13, no. 6 [2004]: 553–82). Since that time, leading educators and institutions have adopted similar strategies, for example, the University of California's Museum of Paleontology project, *Understanding Evolution* (emphasis added). The museum has taken pains to point out that understanding evolution does not need to affect religious belief. See http://undsci.berkeley.edu/article/science_religion (accessed March 26, 2015). Some scholars have offered slightly different approaches. Recently, for example, James D. Williams has called for "acceptance" of evolution, not "belief," as the proper goal of science education. See James D. Williams, "Evolution vs. Creationism, a Matter of Acceptance versus

Belief," *Journal of Biological Education* (2014), http://dx.doi.org/10.1080/00219266.2014.943790 (accessed March 26, 2015), as well as Mike U. Smith and Harvey Siegel, "On the Relationship between Belief and Acceptance as Goals of Evolution Education: Ten Years Later," *Cultural Studies of Science Education* 2015 (forthcoming). Political scientists Michael Berkman and Eric Plutzer, on the other hand, have denounced teachers' understanding-not-belief approaches as a false solution that "seriously undermines the legitimacy of science." See Berkman and Plutzer, *Evolution, Creationism, and the Battle to Control America's Classrooms*, 134.

8. Harry Brighouse, *On Education* (New York: Routledge, 2006); Israel Scheffler, *Reason and Teaching* (Indianapolis: Hackett, 1989); Harvey Siegel, *Educating Reason: Rationality, Critical Thinking, and Education* (London: Routledge, 1988); and Siegel, *Rationality Redeemed?: Further Dialogues on an Educational Ideal* (New York: Routledge, 1997).

9. Such a case was reported in a manuscript submitted for review to a science education journal.

10. The possibility is not overly fanciful; cf. "Texas: Don't Discriminate against Creationists!," *Week*, March 21, 2011, http://theweek.com/article/index/213346/texas-dont-discriminate-against-creationists#axzz33b2JBn33 (accessed April 13, 2015).

11. For further consideration of the non-ET-believing physician, see Smith and Siegel, "On the Relationship between Belief and Acceptance."

12. New Hampshire House Bill 542, June 22, 2011, http://www.gencourt.state.nh.us/legislation/2011/HB0542.html (accessed March 26, 2015), and "Veto Override Gives Parents More Control over What Their Kids Are Taught," *New Hampshire Union Leader* (Manchester, NH), January 4, 2012, http://www.unionleader.com/article/20120104/NEWS04/120109976/0/news05 (accessed March 26, 2015). New Hampshire is not alone in this overreach. In 2012, Missouri voters approved a constitutional amendment giving students and families the right to opt out of offensive school assignments and activities. Other states have considered similar laws and regulations. See Adam Laats, "The Real Wall of Separation in Public Schools," *Washington Post Answer Sheet*, September 9, 2012, http://www.washingtonpost.com/blogs/answer-sheet/post/the-real-wall-of-separation-in-public-schools/2012/09/05/aaf3d766-f6b7-11e1-8253-3f495ae70650_blog.html (accessed March 26, 2015). See also Glenn Branch and Eugenie Scott, "Overcoming Obstacles to Evolution Education: The OOPSIE Compromise—A Big Mistake," *Evolution: Education and Outreach* 1, no. 2 (April 2008): 147–49.

INDEX